CAMBRIDGE LIBRARY COLLECTION

*Books of enduring scholarly value*

## Life Sciences

Until the nineteenth century, the various subjects now known as the life sciences were regarded either as arcane studies which had little impact on ordinary daily life, or as a genteel hobby for the leisured classes. The increasing academic rigour and systematisation brought to the study of botany, zoology and other disciplines, and their adoption in university curricula, are reflected in the books reissued in this series.

## The Fruit Manual

Robert Hogg (1818–97) was a British nurseryman and an early secretary of the Royal Horticultural Society: a prize medal is named in his honour. Born in Berwickshire, Hogg trained in medicine at Edinburgh before following his father into fruit tree cultivation, and became joint editor of the *Cottage Gardener*, later the *Journal of Horticulture*. In 1851, he published *The British Pomology* (also reissued in this series): this work, on apples, was apparently intended as a study of British fruit trees, but no further volumes followed. Instead, in 1860, Hogg published this comprehensive catalogue of British fruit, which ran to five, increasingly extended, editions over the next twenty-five years. It became the standard reference work, and was even plagiarised in *Scott's Orchardist*: however Hogg sued and obtained an injunction preventing further sales. Hogg promoted systematic work in the Royal Horticultural Society and was instrumental in setting up its fruit committee.

Cambridge University Press has long been a pioneer in the reissuing of out-of-print titles from its own backlist, producing digital reprints of books that are still sought after by scholars and students but could not be reprinted economically using traditional technology. The Cambridge Library Collection extends this activity to a wider range of books which are still of importance to researchers and professionals, either for the source material they contain, or as landmarks in the history of their academic discipline.

Drawing from the world-renowned collections in the Cambridge University Library, and guided by the advice of experts in each subject area, Cambridge University Press is using state-of-the-art scanning machines in its own Printing House to capture the content of each book selected for inclusion. The files are processed to give a consistently clear, crisp image, and the books finished to the high quality standard for which the Press is recognised around the world. The latest print-on-demand technology ensures that the books will remain available indefinitely, and that orders for single or multiple copies can quickly be supplied.

The Cambridge Library Collection will bring back to life books of enduring scholarly value (including out-of-copyright works originally issued by other publishers) across a wide range of disciplines in the humanities and social sciences and in science and technology.

# The Fruit Manual

*Containing the Descriptions and Synonymes of the Fruits and Fruit Trees Commonly Met with in the Gardens and Orchards of Great Britain*

Robert Hogg

CAMBRIDGE UNIVERSITY PRESS

Cambridge, New York, Melbourne, Madrid, Cape Town,
Singapore, São Paolo, Delhi, Tokyo, Mexico City

Published in the United States of America by Cambridge University Press, New York

www.cambridge.org
Information on this title: www.cambridge.org/9781108039451



This edition first published 1860
This digitally printed version 2012

ISBN 978-1-108-03945-1 Paperback

# THE FRUIT MANUAL;

CONTAINING

THE DESCRIPTIONS & SYNONYMES

OF

THE FRUITS AND FRUIT TREES

COMMONLY MET WITH IN THE

GARDENS & ORCHARDS OF GREAT BRITAIN,

WITH

*SELECTED LISTS of THOSE MOST WORTHY of CULTIVATION.*

By ROBERT HOGG, LL.D., F.H.S.,

SECRETARY TO THE FRUIT COMMITTEE OF THE HORTICULTURAL SOCIETY OF LONDON, AUTHOR OF "BRITISH POMOLOGY," "THE VEGETABLE KINGDOM AND ITS PRODUCTS," AND CO-EDITOR OF "THE COTTAGE GARDENER."

LONDON:

COTTAGE GARDENER OFFICE,

162, FLEET STREET, E.C.

MDCCCLX.

# PREFACE.

FIFTEEN years ago I published a Manual of Fruits, which at the time included most of the varieties found in nurseries and private gardens. This being favourably received, the whole impression was sold within a twelvemonth, and I was repeatedly urged to prepare a new edition.

About that time numerous new varieties of fruits were introduced to British gardens, and it was therefore necessary that their merits should be fairly tested before a new edition could be published of a work professing to furnish information respecting the fruits and fruit trees commonly cultivated in this country.

During the interval that has elapsed I have examined the greater number of the new, and many of the older varieties not formerly included, and I am now enabled to present a work more complete and useful than I could have done had I entered upon it at an earlier period.

In the present volume I have not attempted to enumerate all the varieties of fruits known to exist in the country, but to describe those only which either are in cultivation, or are worthy of being cultivated for their superior merits. In some instances there are sorts mentioned, not because of their excellence, but because of their popularity from long usage, and in such cases I have stated their true character in comparison with others.

My object has been to prepare a convenient manual of reference for amateur fruit-growers, nurserymen, and professional gardeners, and to condense in a space as small as possible all useful information respecting the varieties of fruits mentioned. I have been particularly careful in regard of the synonymes; and at the end of each of the kinds of fruits I have given selections of varieties for limited gardens, and for different situations and aspects. In most cases I have given a synoptical arrangement of the different fruits by which to facilitate their identification; and I trust that the pains which have been bestowed upon the work generally, will secure for it a favourable reception, and an indulgent consideration for any errors that inadvertently may have occurred.

ROBERT HOGG.

61, WINCHESTER STREET, PIMLICO.

*Sept. 1st*, 1860.

A

# MANUAL OF BRITISH FRUITS.

## APPLES.

[D. signifies that varieties so marked are to be used only for the dessert; K., for kitchen purposes; and C., for cider-making. Those marked K.D. are applicable either to kitchen or dessert use.]

ADAMS' PEARMAIN, D.—Large and pearmain-shaped. Skin pale greenish yellow, tinged and streaked with red on the side next the sun. Eye open. Stalk half an inch long, obliquely inserted. Flesh yellowish, crisp, juicy, and sugary, with a pleasant perfumed flavour. A very handsome and excellent dessert apple. Ripe from December to February.

Alexander. See *Emperor Alexander*.

ALFRISTON, K. (*Lord Gwydyr's Newtown Pippin, Oldaker's New*). — Large, roundish, and irregularly ribbed. Skin light orange next the sun, greenish yellow in the shade, reticulated with russet. Stalk short and deeply inserted. Eye open, set in a deep basin. Flesh yellowish white, crisp, sugary, and sharply acid. A good bearer, and one of the best kitchen apples. November to April.

American Plate. See *Golden Pippin*.

Aporta. See *Emperor Alexander*.

Arbroath Pippin. See *Oslin*.

Arley. See *Wyken Pippin*.

AROMATIC RUSSET, D. (*Brown Spice, Burntisland Pippin, Rook's Nest, Spice Apple*). — Medium sized, conical, flattened at the ends. Skin green and russety. Eye small. Flesh greenish white, richly aromatic. An excellent bearer. October.

ASHMEAD'S KERNEL, D.—Medium sized, roundish, and compressed. Skin greenish yellow and russety, tinged with brown next the sun. Eye small. Stalk short, and deeply inserted. Flesh yellowish, firm, crisp, juicy, sugary, and richly flavoured; of first-rate quality, extensively cultivated near Gloucester, of which neighbourhood it is a native, and well deserving of more general distribution. November to May.

Astrachan. See *White Astrachan.*

Balgone Pippin. See *Golden Pippin.*

BARCELONA PEARMAIN, D. (*Speckled Pearmain, Speckled Golden Reinette*).—Medium sized, oval. Skin yellow in the shade, and a beautiful red next the sun, covered with large russety specks. Stalk short. Eye small. Flesh yellowish, highly aromatic. Abundant bearer, and good dessert apple. November to March.

Bay. See *Drap d'Or.*

Bayfordbury Pippin. See *Golden Pippin.*

BEACHAMWELL, D. (*Motteux' Seedling*). — A small, ovate, yellow apple, of first-rate quality. December to March.

BEAUTY OF KENT, K.—Very large, roundish, flat and russety at the base. Skin greenish yellow, streaked with beautiful red next the sun. Stalk short, slender, and deeply inserted. Eye small. Flesh crisp, tender, and juicy. An abundant bearer. October to February.

BEDFORDSHIRE FOUNDLING, K. (*Cambridge Pippin*).—Large, roundish, and slightly ribbed. Skin dark green, becoming paler as it ripens. Stalk short, and deeply inserted. Eye open and deep. Flesh yellowish, and pleasantly acid. Handsome and excellent. November to March.

Bell's Scarlet. See *Scarlet Pearmain.*

BESS POOL, K.D.—Above medium size, conical, and handsomely shaped. Skin yellow, washed and striped with red on the side next the sun. Eye small, and rather deep. Stalk short. Flesh white, tender, and juicy, with a fine sugary and vinous flavour. November to March.

BLENHEIM ORANGE, K.D. (*Woodstock Pippin, Northwick Pippin*).—Large, round, and widest at the base. Skin yellowish, red next the sun. Eye open and hollow. Flesh yellow, sweet, and juicy. A first-rate dessert fruit, and excellent for kitchen use. November to February.

BOROVITSKI, D.—Medium sized, roundish, and angular.

Skin bright red on one side and pale green on the other. Stalk long and deeply inserted. Flesh white, brisk, juicy, and sugary. Middle of August.

BORSDÖRFFER, D. (*Garret Pippin, King George, Queen's*).—Small, oval, bright yellow, and red next the sun. Stalk short and slender. Eye shallow. Flesh yellowish white, crisp, rich, and perfumed. An apple of very superior quality. November to March.

BOSTON RUSSET, D. (*Roxbury Russet*).—Medium sized, roundish, flattened at the ends. Skin dull green, covered with brownish-yellow russet, rarely tinged with red. Stalk nearly an inch long, slender. Flesh greenish white, rich, sub-acid, and juicy, like Ribston Pippin. Of first-rate quality. January to April.

BRABANT BELLEFLEUR, K.D.—Large, roundish, oblong, and ribbed. Skin pale yellow, slightly striped with red. Eye large and wide. Flesh juicy, rich, and pleasantly sub-acid. November to April.

BRADDICK'S NONPAREIL, D. (*Ditton Nonpareil*).—Small, roundish, and compressed at both ends. Skin smooth, green, tinged with yellowish brown, brownish red next the sun. Eye small and deeply set. Stalk short. Flesh yellow, sugary, and aromatic. An abundant bearer, and first-rate table fruit. December to March.

Brandy. See *Golden Harvey*.

BRINGEWOOD PIPPIN, D.—Small and round. Skin of a fine rich yellow colour. Eye small and open, set in a shallow basin. Stalk short and slender. Flesh yellowish, firm, crisp, juicy, and with a rich flavour. A first-rate dessert apple. January to March.

BROWNLEES' RUSSET, K. D.—Large, roundish-ovate, and rather flattened. Skin green and russeted, with brownish red next the sun. Eye closed. Stalk short, deeply inserted. Flesh greenish white, tender, juicy, sweet, and aromatic. An excellent late apple. January to May.

Brown Spice. See *Aromatic Russet*.

Burntisland Pippin. See *Aromatic Russet*.

CALVILLE, WINTER WHITE, K. (*White Calville*).—Large and flattened, marked on its sides with prominent ribs. Skin smooth, shining, rich yellow, and tinged with red. Eye small and deep. Stalk slender and deeply inserted. Flesh white, tender, sweet, and juicy. January to April.

Cambridge Pippin. See *Bedfordshire Foundling*.

Carel's Seedling. See *Pinner Seedling*.

CARLISLE CODLIN, K.—Fruit above medium size, ovate, and angular. Skin smooth, pale yellow. Eye closed. Stalk very short. Flesh white, tender, crisp, juicy, and brisk. An excellent culinary apple. From August to December.

CELLINI, K. D.—Above medium size, roundish, and handsomely shaped. Skin deep yellow, beautifully streaked and mottled with red next the sun. Eye large and open, set in a shallow basin. Stalk very short. Flesh white, tender, juicy, with a fine, brisk, balsamic flavour, and high aroma. A first-rate culinary apple, and also useful in the dessert. October to November.

Chalmers' Large. See *Dutch Codlin*.

CHRISTIE'S PIPPIN, D.—Rather small, round, and compressed. Skin deep yellow, mottled with red next the sun. Stalk short. Eye small. Flesh yellowish white, tender, brisk, and juicy, with a pleasant flavour. A first-rate dessert apple. December to February.

Claremont. See *French Crab*.

CLAYGATE PEARMAIN, D.—Medium sized, conical. Skin dull greenish yellow, with brownish red next the sun. Stalk medium. Eye large. Flesh yellow, tender, and aromatic, with the flavour of the Ribston. An abundant bearer, and first-rate fruit. November to March.

Clifton Nonesuch. See *Fearn's Pippin*.

Coates'. See *Yorkshire Greening*.

Cobbett's Fall Pippin. See *Reinette Blanche d'Espagne*.

COBHAM, D. — Above medium size, roundish. Skin greenish yellow, mottled with red. Eye small and closed. Stalk slender and deeply inserted. Flesh pale yellow, crisp, sugary, and aromatic. An excellent dessert apple, with something of the character of Ribston Pippin. September to January.

COCCAGEE, C.—Medium sized, ovate, fine yellow. Skin red next the sun. One of the best cider apples. October to December.

COCKLE PIPPIN, D. (*Nutmeg Pippin*).—Medium sized, conical or ovate. Skin fine brownish yellow, russety at the base. Stalk slender. Eye in a narrow and shallow basin. Flesh yellow, rich, and perfumed. Excellent flavour, and first-rate dessert fruit. January to April.

COE'S GOLDEN DROP, D.—Small and conical. Skin yellow, with a few crimson spots next the sun. Eye small and open. Stalk long. Flesh firm, crisp, sugary, and vinous. A first-rate dessert apple. November to May.

Copmanthorpe Crab. See *Dutch Mignonne*.

CORNISH GILLIFLOWER, D.—Rather large, oval, and angular towards the eye. Skin deep yellowish green, tinged with red, intermixed with streaks of deeper red next the sun, russety. Stalk an inch long. Eye nearly closed, set in an uneven basin. Flesh yellow, firm, rich, and perfumed, like the Clove Gilliflower. Rather a shy bearer, but one of "the best of apples." November to May.

COURT OF WICK, D. (*Fry's Pippin, Golden Drop, Knightwick Pippin, Phillips' Reinette, Wood's Huntingdon, Weeks' Pippin, Yellow Pippin*).—Rather small, roundish, ovate, and compressed at the ends. Skin greenish yellow, orange and russety next the sun. Stalk short and slender. Eye open and shallow. Flesh deep yellow, juicy, and highly flavoured. An abundant bearer, and first-rate fruit. October to March.

COURT PENDU PLAT, D. (*Garnon's Pippin, Princesse Noble Zoete, Russian, Wollaton Pippin*)—Medium sized, round, and compressed. Skin rich deep red, greenish yellow in the shade. Stalk short and deeply inserted. Eye large and open, set in a wide shallow basin. Flesh yellow, rich, and briskly acid. An abundant bearer, and excellent fruit. November to April.

COX'S ORANGE PIPPIN, D.—Medium sized, roundish-ovate, and regular in its outline. Skin greenish yellow, and streaked with red in the shade, but dark red where exposed to the sun. Eye small and open. Stalk half an inch long. Flesh yellowish, very tender, crisp, and juicy, with a fine perfume. A first-rate dessert apple. October to February.

COX'S POMONA, K.—Above the medium size, sometimes large, ovate, and somewhat flattened and angular. Skin yellow, and very much streaked with bright crimson. Eye slightly open and deep. Stalk an inch long, deeply inserted. Flesh white, tender, delicate, and pleasantly acid. October.

Crofton Scarlet, d.—Medium sized, flattish. Skin yellowish russet, bright red and russety next the sun. Eye wide. Stalk short. An abundant bearer. October to December.

Devonshire Quarrenden, d. (*Red Quarrenden, Sack Apple*).—Medium sized, round, compressed at the ends. Skin deep crimson. Stalk short and deeply inserted. Eye with long segments, very shallow. Flesh greenish white, crisp, juicy, and pleasantly sub-acid. A good bearer. "No better autumn fruit." August.

Devonshire Queen, k.d.—A fine, large, ovate fruit, entirely covered with rich, dark crimson, and a delicate bloom. The flesh is sometimes tinged with red, and is crisp, juicy, and balsamic. October.

Ditton Nonpareil. See *Braddick's Nonpareil.*

Downton Pippin, d.—Larger than the Golden Pippin, roundish, flat at the ends. Skin yellow. Stalk short and deeply inserted. Eye in a wide and shallow basin. Flesh yellow, brisk, and richly flavoured. A seedling from the Golden Pippin. November to January.

Drap d'Or, k. (*Bay Apple, Early Summer Pippin*).—Rather large, roundish, narrowing towards the eye. Skin yellow, dotted with brown specks. Stalk short. Eye shallow. Flesh crisp, juicy, and of a pleasant mild flavour. October to December.

Duchess of Oldenburgh, d.—Medium sized, roundish. Skin rich yellow, streaked with red. Eye large, nearly closed, set in a wide hollow. Flesh brisk and juicy. September.

Dumelow's Seedling, k. (*Normanton Wonder, Wellington*).—Large, round, and compressed at both ends. Skin yellow, light red next the sun. Stalk very short. Eye large and open. Flesh yellow. A good bearer, and an excellent kitchen apple. November to March.

Dundee. See *Golden Reinette.*

Dutch Codlin, k. (*Chalmers' Large, Glory of the West*).—Very large, irregularly roundish, or oblong, with prominent ribs extending from the base to the eye. Skin pale greenish-yellow, slightly tinged with orange, red next the sun. Stalk short and thick. Eye set in a deep angular basin. Flesh white, slightly acid. A good bearer, and one of the best kitchen apples. August to September.

Dutch Mignonne, K.D. (*Copmanthorpe Crab, Stettin Pippin*).—Rather large, roundish, and handsome. Skin dull orange, half mottled with large yellow russet specks. Eye open, deeply set in a round basin. Stalk an inch long, deeply set. Flesh highly aromatic. A great bearer, and one of the most desirable apples for any garden. December to April.

Early Crofton. See *Irish Peach*.

Early Harvest, D. (*Yellow Harvest*).—Medium sized, round. Skin clear pale yellow. Eye small and closed. Stalk half an inch long, not deeply inserted. Flesh white, tender, crisp, juicy, with a pleasant refreshing flavour. A first-rate early dessert apple. July and August.

Early Julien, K.D.—Medium sized, roundish, and slightly flattened. Skin pale yellow, with an orange tinge next the sun. Eye closed. Stalk short. Flesh yellowish white, crisp, very juicy, with a fine brisk and rather balsamic flavour. An excellent early apple. Ripe in the second week of August.

Early Nonpareil, D. (*Hicks' Fancy, New Nonpareil, Stagg's Nonpareil*).—Small, roundish, narrowing towards the eye. Skin greenish yellow, changing to deep yellow as it attains maturity, russety, and spotted with grey spots. Eye open, set in a wide basin. Stalk short and deeply inserted. Flesh yellowish white, crisp, juicy, brisk, and aromatic. October to December.

Early Red Margaret. See *Margaret*.

Early Summer Pippin. See *Drap d'Or*.

Easter Pippin. See *French Crab*.

Edmonton Aromatic. See *Kerry Pippin*.

Elizabeth. See *Golden Reinette*.

Emperor Alexander, K. (*Aporta, Russian Emperor*).—Very large, heart-shaped. Skin greenish yellow, streaked with bright red next the sun. Eye large and deeply set. Stalk slender, an inch long, much inserted. Flesh yellowish white, rich, juicy, and aromatic. A very handsome apple. September to December.

English Codlin, K.—Large, conical, and irregular in

its outline. Skin fine yellow, with a faint red blush on the side exposed to the sun. Eye closed. Stalk short and stout. Flesh white, tender, and agreeably acid. August to October.

English Pippin. See *Golden Reinette.*

Fall Pippin. See *Reinette Blanche d'Espagne.*

FEARN'S PIPPIN, K.D. (*Ferris' Pippin, Clifton Nonesuch*). —Medium sized, round, flat at the ends. Skin greenish yellow, russety round the stalk, and bright red next the sun. Stalk short. Eye shallow, in a plaited basin. Flesh greenish white, sweet, and richly flavoured. A good apple. November to February.

FEDERAL PEARMAIN, D.—Below medium size, pearmain-shaped. Skin yellowish, with a little red, and a few dark streaks on the side next the sun, russety. Eye deeply set. Stalk half an inch long. Flesh fine, delicate, very juicy, and of excellent flavour. A first-rate apple. December to March.

Five-crowned Pippin. See *London Pippin.*

FLOWER OF KENT, K.—Large, roundish, flattened, and irregularly ribbed. Skin dull yellow, tinged with red, bright red next the sun. Flesh greenish yellow, exceedingly juicy. Eye small. Stalk an inch long. October to January.

FORGE, K.—Medium sized, round. Skin a golden-yellow colour, mottled with crimson, and dark red next the sun. Eye small and closed. Stalk very short. Flesh yellowish white, tender, juicy, sweet, and finely perfumed. A useful apple. The tree a great and constant bearer. October to January.

FORMAN'S CREW, D.—Below medium, oval, broadest at the base. Skin yellow and russety. Stalk short. Eye small. Flesh greenish yellow. One of the best dessert apples. November to April.

Formosa. See *Ribston Pippin.*

FRANKLIN'S GOLDEN PIPPIN, D. (*Sudlow's Fall*).—Medium size, conical. Skin bright yellow, dotted with dark spots. Stalk short, slender, and deeply set. Eye deeply sunk. Flesh pale yellow, tender, and richly aromatic. A first-rate fruit. October to January.

FRENCH CRAB, K.D. (*Claremont Pippin, Easter Pippin, Ironstone, Young's Long Keeping*). — Large, globular.

Skin dark green, with a brown blush next the sun. Stalk short and slender, deeply set. Eye small, almost closed. Flesh pale green, firm, and pleasantly sub-acid. An immense bearer, and remarkable for keeping, under favourable circumstances, for two years.

Frith Pitcher. See *Manks Codlin.*

Fry's Pippin. See *Court of Wick.*

Garnon's. See *Court Pendu Plat.*

Garret Pippin. See *Borsdörffer.*

Girkin Pippin. See *Wyken.*

GLORIA MUNDI, K. (*Baltimore, Mammoth, Ox Apple, Monstrous Pippin.*—Of very large size, roundish, and flattened. Skin pale yellowish green, with a faint tinge of blush on one side. Eye large and open. Stalk short and stout. Flesh white, tender, and juicy. October to Christmas.

GOLDEN KNOB. D.—Small, ovate, and a little flattened. Skin yellow, much covered with russet, with a reddish tinge on one side. Eye open, stalk very short. Flesh greenish white, firm, crisp, and juicy. December to March.

Glory of the West. See *Dutch Codlin.*

Glory of York. See *Ribston Pippin.*

Golden Drop. See *Court of Wick.*

GOLDEN HARVEY, D.C. (*Brandy*).—Small, nearly round. Skin roughly russety, on a yellow ground, tinged with red next the sun. Stalk half an inch long, slender. Eye small, open, and shallow. Flesh yellow, rich, aromatic, and sub-acid flavour. A first-rate dessert fruit. December to June.

GOLDEN MONDAY, D.—Small, roundish, and flattened. Skin clear, golden yellow, with markings of russet. Eye small, and rather open. Stalk very short. Flesh yellowish white, crisp, sugary, briskly flavoured, and with a nice aroma. October to Christmas.

GOLDEN NOBLE, K. (*Waltham Abbey Seedling*).—A very large, globular, and handsome apple. Skin of a uniform clear, bright yellow. Eye small and deep. Stalk short. Flesh yellow, tender, and pleasantly acid, and bakes of a clear amber colour. A valuable kitchen apple. September to December.

Golden Pippin, d. (*American Plate, Balgone Pippin, Bayfordbury Pippin, Herefordshire G.P., London G.P., Melton G.P., Russet G.P., Warter's G.P.*) — Small, roundish. Skin deep golden yellow, with white specks under the skin, dotted with russet. Stalk long and slender. Eye small and shallow. Flesh yellowish, rich, brisk, and highly flavoured. The queen of dessert apples. November to March.

Golden Reinette, d. (*Dundee, English Pippin, Elizabeth, Kirke's Golden Reinette, Megginch Favourite, Princess Noble, Wyker Pippin, Wygers*).—Below medium size, round, and compressed at the ends. Skin greenish yellow, flushed and streaked with red next the sun, dotted with russet. Stalk long. Eye large, open, and shallow. Flesh yellow, sugary, and richly flavoured. October to January.

Golden Russet, d. — Medium sized, ovate. Skin greenish yellow, covered with yellow russet. Stalk short. Eye small and close. Flesh yellowish white, crisp, and pleasantly aromatic. December to March.

Golden Winter Pearmain, k.d. (*King of the Pippins, Hampshire Yellow, Jones' Southampton Pippin*).—Medium sized, abrupt pearmain-shaped. Skin rich yellow, tinged and streaked with red next the sun. Eye large and open, set in a deep basin. Stalk long and stout. Flesh yellowish white, firm, juicy, and sweet, with a somewhat aromatic flavour. A valuable apple. October to January.

Gooseberry Pippin, k. — Medium sized, roundish. Skin deep lively green. Eye open. Stalk short. Flesh greenish white, very tender, and delicate. A very valuable, late-keeping, culinary apple. In use from November till the following August.

Gravenstein, k.d. — Large, round, flattened at the ends, and angular. Skin fine straw colour, streaked with red next the sun. Stalk very short and deeply set. Eye large, wide, and deeply set. Flesh pale yellow, crisp, aromatic, and vinous. A very valuable apple. October to December.

Greenup's Pippin, k.—Above medium size, round. Skin of a pale straw colour, with a fine bright red cheek next the sun. Eye closed. Stalk short. Flesh pale yellowish white, juicy, sweet, and brisk. October to December.

Grey Leadington, D. — Medium sized, oblong or conical, and ribbed. Skin yellow and russety, with pale red on the side exposed to the sun. Eye large and sunk. Stalk short and very stout. Flesh tender, juicy, sugary, and finely perfumed. An excellent dessert apple. In use from September to January.

Hall Door, D. — Medium sized, roundish. Skin greenish yellow in the shade, and streaked with red on the side next the sun. Eye small, and set in a rather deep basin. Stalk short, and inserted in a deep cavity. Flesh white, firm, and juicy. In use from November to March.

Hambledon Deux Ans, K.D.—Large roundish, rather broadest at the base. Skin yellowish green in the shade, and dull-red, streaked with broad stripes of a deeper red, on the side next the sun. Eye small and closed. Stalk stout and short. Flesh greenish white, firm, crisp, and richly flavoured. One of the most valuable keeping apples. In use from January to May.

Hampshire Yellow. See *Golden Winter Pearmain.*

Hanwell Souring, K.—Medium sized, roundish-ovate. Skin greenish yellow, with a red blush. Eye closed. Stalk very short. Flesh firm, crisp, and briskly acid. Worthy of general cultivation. December to March.

Harvey Apple, K. — Large, roundish-ovate. Skin greenish yellow, with markings of russet. Eye small. Stalk short and slender. Flesh white, crisp, juicy, and pleasantly acid. A first-rate culinary apple. October to January.

Hardingham's Russet. See *Pine Apple Russet.*

Hawberry Pippin. See *Hollandbury.*

Hawthornden, K.—Large, flat, ovate, and angular. Skin delicate yellowish-green, covered with bloom, a red blush next the sun. Stalk slender, half an inch long. Eye small, nearly closed. Flesh white, juicy, and pleasant. An abundant bearer. September to January.

Hawthornden, New, K.—The appearance of the fruit is very much the same as that of the old Hawthornden, but is much more solid and briskly flavoured. It also keeps longer, and the tree has a more robust and vigorous growth. December to January.

Herefordshire Golden Pippin. See *Golden Pippin.*

Herefordshire Pearmain. See *Royal Pearmain.*

Hicks' Fancy. See *Early Nonpareil.*

HOARY MORNING, K. (*Dainty, Downy, Sam Rawlings*).—Large, roundish, somewhat flattened, and angular. Skin yellowish, marked with broad, pale-red stripes on the shaded side, and broad broken stripes of beautiful red on the side next the sun, and covered with a fine thick bloom, like thin hoar frost. Eye very small. Stalk short. Flesh yellowish white, tinged with red at the surface under the skin, brisk, juicy, rich, and slightly acid. This is a beautiful and very excellent kitchen apple. In use from October to December.

HOLBERT'S VICTORIA, D. — Small and ovate. Skin yellow, covered with pale grey russet Eye small and slightly open. Stalk short. Flesh yellowish, firm, very juicy, vinous, and aromatic. An excellent dessert apple, of the first quality. December to May.

HOLLANDBURY, K. (*Hawberry Pippin, Horsley Pippin, Kirke's Admirable*).—Large, roundish, flat at the ends, prominently ribbed. Skin greenish yellow, beautiful bright red next the sun. November to January.

HOLLAND PIPPIN, K.—Large, roundish, and flattened. Skin yellow, inclining to green, dull red next the sun. Stalk short, thick, and deeply set. Eye small, in a slightly plaited basin. Flesh pale yellow and pleasantly acid. November to March.

Hood's Seedling. See *Scarlet Pearmain.*

HORMEAD PEARMAIN, K.D. (*Arundel Pearmain, Hormead Pippin*).—Medium sized, ovate conical. Skin of a uniform bright yellow. Eye large, closed, and set in a shallow, irregular basin. Stalk very short and stout, inserted in a deep cavity. Flesh white, tender, very juicy, and pleasantly acid. An excellent dessert apple. In use from October to March.

Horsley Pippin. See *Hollandbury.*

HUBBARD'S PEARMAIN, D.—Small and conical. Skin covered with thin russet, sometimes without russet, and thin yellowish green. Eye small and closed. Stalk short. Flesh yellow, firm, sugary, richly flavoured, and aromatic. One of the best dessert apples, deserving extensive cultivation. November to April.

HUGHES' GOLDEN PIPPIN, D. — Small, round, compressed at the ends. Skin yellow, spotted with green, and russety. Stalk thick and short. Eye small, in a hollow, plaited basin. Flesh yellow, rich, sweet, and agreeable. A first-rate dessert apple. December to February.

Hunt's Nonpareil. See *Nonpareil.*

INGESTRIE RED, D.—Small, oblong ovate. Skin bright yellow, tinged with red next the sun, speckled with dots. Eye in a round, wide basin. Stalk short and slender. Flesh pale yellow, of very rich, juicy flavour, resembling the Golden Pippin. One of the best autumn apples. September to November.

INGESTRIE YELLOW, D.—Small, ovate oblong. Skin deep bright yellow all over. Eye small and shallow. Stalk slender. Flesh tender, and very juicy when first gathered. A good bearer. October.

IRISH PEACH, D. (*Early Crofton*).—Medium sized, roundish, somewhat flattened and angular. Skin yellowish green and dotted with brown in the shade, dull red next the sun. Eye large and closed. Stalk short. Flesh white, tender, juicy, and richly flavoured. This is one of the best summer dessert apples, and is ripe in August.

Irish Pitcher. See *Manks Codlin.*

Irish Russet. See *Sam Young.*

Ironstone. See *French Crab.*

ISLE OF WIGHT PIPPIN, K.D.—Medium sized, round, and a little flattened. Skin fine rich yellow, and covered with thin grey russet on the shaded side, and of a rich orange and red on the side next the sun. Eye open. Stalk short. Flesh of a fine yellow colour, firm, juicy, and pleasantly acid. September to January.

JOANNETING, D. (*Juneating, White Juneating, Owens' Golden Beauty*).—Small, round, and slightly flattened. Skin light yellow, with a red blush next the sun. Eye moderately sunk. Stalk long and slender. Flesh crisp and pleasant. A good bearer. July to August.

Jones' Southampton Pippin. See *Golden Winter Pearmain.*

KEDDLESTON PIPPIN, D.—Small, conical, and regularly formed. Skin of a uniform yellow colour, with veinings and specks of russet. Eye half open, set in a shallow, plaited basin. Stalk short. Flesh yellowish, crisp, very juicy, sugary, and aromatic. A first-rate dessert apple. November to March.

KENTISH FILL BASKET, K. (*Lady De Grey's, Potters' Large*).—Very large, roundish, and angular. Skin yellowish green, with a brownish red blush next the sun, streaked with darker red. Eye large, in a deep irregular basin. Flesh juicy, sub-acid. November to January.

KERRY PIPPIN, D. (*Edmonton Aromatic Pippin*).—Medium sized, oval, flattened and wrinkled at the eye. Skin pale yellow, tinged and streaked with red next the sun. Eye obliquely inserted in a plaited basin. Stalk large. Flesh yellow, firm, crisp, and very juicy, with a rich sugary flavour. One of the best dessert apples. September to October.

KESWICK CODLIN, K.—Large, conical, irregularly angular. Skin greenish yellow, with a blush tinge next the sun. Eye large, deeply set. Stalk short, much depressed. Flesh yellowish white, juicy, and pleasantly sub-acid. An excellent bearer. August to September.

King George. See *Borsdörffer.*

King of the Pippins. See *Golden Winter Pearmain.*

Kirke's Admirable. See *Hollandbury.*

Kirke's Golden Reinette. See *Golden Reinette.*

Kirke's Lemon Pippin. See *Lemon Pippin.*

KIRKE'S LORD NELSON, K.—Large, roundish. Skin smooth, pale yellow, streaked all over with red. Eye open. Stalk short and slender. Flesh yellowish white, sweet and juicy, but lacks acidity. November to February.

Knight's Codlin. See *Wormsley Pippin.*

Knightwick. See *Court of Wick.*

Lady De Grey's. See *Kentish Fill Basket.*

LAMB ABBEY PEARMAIN, D.—Medium sized, conical, slightly flattened at the ends. Skin yellowish green; orange, streaked with red, next the sun. Eye rather large, deeply sunk. Stalk short. Flesh yellow, greenish at the core, crisp, juicy, sweet, and aromatic. One of the best dessert fruits, and "keeps well without shrivelling." December to April.

Lancashire Crab. See *Minchall Crab.*

Leathercoat. See *Royal Russet.*

LEMON PIPPIN, K.D. (*Kirke's Lemon Pippin*).—Medium sized, oval. Skin yellowish green, turning to lemon-yellow. Eye small. Stalk short, with a fleshy protuberance growing on one side. Flesh firm, brisk, and pleasantly acid. October to April.

LEWIS' INCOMPARABLE, K.D.—Large and conical. Skin deep lively red, streaked with red of a darker colour; but on the shaded side it is deep yellow, faintly streaked with light red, and strewed with numerous minute dark dots. Eye small and open. Stalk very short. Flesh yellowish, firm, crisp, and juicy, with a slight musky flavour. A large and handsome apple. In use from December to February.

LINCOLNSHIRE HOLLAND PIPPIN, K. (*Striped Holland Pippin*).—Large, roundish, and flattened. Skin yellow, dotted with green dots in the shade, and pale orange, streaked with bright red, next the sun. Eye small, set in an angular basin. Stalk short, inserted in a shallow depression. Flesh white, and slightly acid. A kitchen apple from October to December.

LITTLE HERBERT, D.—A small, round apple, covered with brown russet. Much esteemed in Gloucestershire as a first-rate dessert fruit; but is, in fact, only second-rate. Tree a shy bearer. December to March.

LOAN'S PEARMAIN, D.—Rather large, oval. Skin dull green, with brownish red next the sun. Flesh greenish white, tender, crisp, juicy, and sweet. November to February.

London Golden Pippin. See *Golden Pippin*.

LONDON PIPPIN, K. (*Five-crowned Pippin, Royal Somerset*).—Above medium size, roundish, flattened, and angular, with five prominent knobs round the crown. Skin yellow, becoming deep yellow when ripe. Stalk short and slender. Eye small and shallow. Flesh yellowish white, of an agreeable sub-acid flavour. Good bearer. October to January.

LONGVILLE'S KERNEL, D. (*Sam's Crab*). — Medium sized, ovate, and slightly angular. Skin greenish yellow, streaked with dark yellow on the side next the sun. Eye small and open. Stalk short. Flesh yellow, firm, slightly acid, and sweet. Only a second-rate apple. August and September.

Lord Gwydyr's Newtown Pippin. See *Alfriston*.

Lord Nelson, k.d.—Rather large, roundish, broadest at the base. Skin pale yellow, bright red next the sun. Stalk slender and short. Eye open, set in a large, slightly plaited basin. Flesh yellowish, juicy, highly aromatic. November to January.

Lord Suffield, k.—Above medium size, conical. Skin pale greenish yellow, with sometimes a tinge of red next the sun. Eye closed. Stalk short and stout. Flesh white, tender, and firm, very juicy, and briskly flavoured. August and September.

Lovedon's Pippin. See *Nonpareil.*

Lucombe's Pine, d.—Below medium size, ovate or conical. Skin clear pale yellow, with an orange tinge next the sun, and marked with patches of russet. Eye small and closed. Stalk short and stout. Flesh tender, crisp, very juicy, sugary, and aromatic. A first-rate dessert apple. October to Christmas.

Lucombe's Seedling, k.—Large, roundish, and angular. Skin yellowish green, covered with dark spots, and streaked with crimson next the sun. Stalk short and thick. Eye small and open. Flesh white, juicy, and pleasant flavoured. October to March.

Maclean's Favourite, d.—Medium sized, roundish. Skin yellow. Flesh crisp and richly flavoured, resembling the Newtown Pippin. An abundant bearer, and "of the highest excellence." October to January.

Mammoth. See *Gloria Mundi.*

Manks Codlin, k. (*Irish Pitcher, Frith Pitcher*).—Medium sized, conical. Skin pale yellow, flushed with red next the sun. Stalk short and fleshy. Eye shallow. Flesh yellowish white, slightly perfumed. One of the best kitchen apples. September to November.

Mannington's Pearmain, d.—Medium sized, abrupt pearmain-shaped. Skin rich golden yellow, covered with russet, and dull brownish red next the sun. Eye closed. Stalk long. Flesh yellow, firm, crisp, juicy, sugary, and aromatic. A first-rate late dessert apple. October to March.

Margaret, d. (*Early Red Margaret, Eve, Red Juneating, Striped Juneating, Striped Quarrenden*).—Under the medium size, roundish-ovate, narrowing towards the eye. Skin greenish yellow, with deep red stripes on one side.

Eye small, closed, and set in a shallow plaited basin. Flesh white, brisk, juicy, and vinous. "One of the best early apples," generally eaten off the tree. August.

MARGIL, D. (*Neverfail*).—Small, ovato-conical, slightly angular. Skin orange, streaked and mottled with red next the sun, slightly russety in the shade. Stalk short. Eye small, set in an irregular basin. Flesh yellow, firm, and richly aromatic. November to March.

Megginch Favourite. See *Golden Reinette*.

MELON APPLE, D.—Medium sized, roundish, and narrowing a little towards the eye. Skin lemon yellow on the shaded side, and light crimson next the sun. Eye small and half open. Stalk half an inch long, very slender. Flesh yellowish white, very tender, crisp, juicy, sweet, and vinous, with a delicate perfume. A first-rate American apple, which ripens in this country. December and January.

MERE DE MENAGE, K.—Large, conical. Skin red, streaked with darker red all over, except a little on the shaded side, where it is yellow. Eye sunk in an angular basin. Stalk very stout, inserted in a deep cavity, so much so as to be scarcely visible. Flesh firm, crisp, brisk, and juicy. A valuable and beautiful kitchen apple. In use from October to January.

Milton Golden Pippin. See *Golden Pippin*.

MINCHALL CRAB, K. (*Lancashire Crab*).—Large, round, and considerably depressed. Skin dull green on the shaded side, and tinged and striped with dull red on the side next the sun. Eye large and open. Stalk rather short. Flesh white, firm, crisp, and briskly acid. November to March.

MINIER'S DUMPLING, K.—Large, roundish, somewhat flattened, and angular. Skin dark green, covered with dark red next the sun. Stalk an inch long, and stout. Flesh firm, juicy, and sub-acid. November to May.

Monstrous Pippin. See *Gloria Mundi*.

MORRIS' NONPAREIL RUSSET, D.—Small, conical, and with the eye placed laterally. Skin green, covered with large patches of russet. Eye small and open. Stalk short and deeply inserted. Flesh greenish, firm, crisp, juicy, sugary, and aromatic. An excellent dessert apple. October to March.

Mother Apple. See *Oslin.*

Mother Apple, American, d.—Above medium size, conical and angular. Skin deep yellow, but highly coloured, with veins and mottles of crimson. Eye small and closed. Stalk half an inch long, slender. Flesh yellowish white, vey tender, and juicy; crisp, sweet, and with a balsamic flavour. A first-rate American apple, which ripens well in this country. October.

Motteux' Seedling. See *Beachamwell.*

Nanny, d. — Medium sized, roundish, angular and ribbed round the eye. Skin greenish yellow, streaked with crimson, and often with a deep red cheek. Eye open. Stalk short. Flesh yellow, soft, and tender, juicy, sugary, and with the Ribston Pippin and Margil flavours. A first-rate dessert apple. In use in October, but soon becomes mealy.

Nelson Codlin, k.d. (*Nelson*).—Large and handsome, of a conical or oblong shape. Skin of a uniform rich yellow, and covered with rather large dark dots. Eye open, set in a deep, plaited, irregular basin. Stalk very short. Flesh yellowish white, delicate, tender, juicy, and sugary. A valuable apple for the kitchen or even the dessert. Season from September to January.

Neverfail. See *Margil.*

New Nonpareil. See *Early Nonpareil.*

Newtown Pippin, d. — Medium sized, roundish, rather irregular, and obscurely ribbed. Skin dull green, changing to olive green when ripe, with a brownish blush next the sun. Eye small and closed, set in a shallow basin. Stalk short, slender, and deeply set. Flesh greenish white, tender, juicy, and crisp, with a fine aroma. Requires a wall in this country. December to April.

New York Gloria Mundi. See *Gloria Mundi.*

Nonesuch, k.d.—Medium sized, round, and flattened. Skin greenish yellow, striped with dull red next the sun. Stalk short and slender. Eye small, set in a wide, shallow basin. Flesh white, tender, and pleasantly sub-acid. September and October.

Nonpareil, d. (*Hunt's Nonpareil, Lovedon's Pippin*) —Rather below medium size, roundish, slightly ovate. Skin greenish yellow, with pale russet, and brownish red next the sun. Stalk long and slender. Eye small, set in a narrow, round basin. Flesh greenish white, firm, crisp, and richly flavoured. A first-rate dessert apple. January to May.

Norfolk Bearer, k.—About medium size, roundish, angular round the eye. Skin green, with a yellowish tinge on the shaded side, but covered with dark crimson next the sun. Eye small and slightly open. Stalk half an inch long, slender. Flesh greenish, tender, crisp, with a brisk and agreeable flavour. Tree a great bearer. December and January.

Norfolk Beefing, k. (*Catshead Beefing*, *Read's Baker*).—Large, round, flat at the ends. Skin green, and deep red next the sun. Stalk short, fleshy, and deeply inserted. Eye large, set in a deep and irregularly plaited basin. Flesh greenish white, firm, and sub-acid. "Excellent for drying." November to July.

Norfolk Colman. See *Winter Colman*.

Norfolk Pippin. See *Adams' Pearmain*.

Norfolk Storing. See *Winter Colman*.

Normanton Wonder. See *Dumelow's Seedling*.

Northern Greening, k. (*Walmer Court*).—Above the medium size, roundish ovate. Skin dull green, brownish red next the sun. Stalk short and thick. Eye small. Flesh greenish white, sub-acid. A first-rate kitchen apple. November to April.

Northern Spy, d.—Large, conical, and angular. Skin yellow on the shaded side, but streaked with crimson on the side next the sun. Stalk three quarters of an inch long, slender. Flesh yellowish white, juicy, rich, and aromatic. An American apple, which ripens well in this country. December to May.

Northwick Pippin. See *Blenheim Orange*.

Nutmeg Pippin. See *Cockle Pippin*.

Oldaker's New. See *Alfriston*.

Old Maid's. See *Knobbed Russet*.

Orange Pippin. See *Isle of Wight Pippin*.

Ord's Apple, d.—Medium sized, conical, with prominent and unequal ribs, forming ridges round the eye. Skin smooth and shiny, light green, and with pale brownish red next the sun. Eye small and closed. Stalk short. Flesh tender, crisp, and brittle; very juicy, vinous, and perfumed. An excellent dessert apple. January to May.

Oslin, d. (*Arbroath Pippin, Mother Apple*).—Medium sized, roundish, flattened. Skin green, changing to lemon yellow, dotted with greyish-green specks. Stalk thick and short. Eye in a shallow, plaited basin. Flesh yellowish, firm, crisp, and highly aromatic. One of the best summer apples. August.

Owen's Golden Beauty. See *Joanneting.*

Ox Apple. See *Gloria Mundi.*

Oxford Peach. See *Scarlet Pearmain.*

Pearmain. See *Winter Pearmain.*

Pearson's Plate, d.—Small, roundish, and flattened. Skin greenish yellow, red next the sun. Eye open and shallow. Stalk half an inch long. Flesh greenish yellow, firm, crisp, juicy, and sugary, with a fine brisk flavour. A first-rate dessert fruit. December to March.

Pennington's Seedling, d.—Medium sized, flat, and slightly angular. Skin covered with yellow russet, pale brown next the sun. Stalk long, thick, and set in a wide, irregular cavity. Eye with long segments, shallow. Flesh yellowish, firm, crisp, and highly flavoured. November to March.

Phillipps' Reinette. See *Court of Wick.*

Piles' Russet, d.—Rather large, irregular. Skin pale green, and covered with thick russet. Stalk short. Eye closed. Flesh greenish yellow, firm, sugary, and aromatic. March and April.

Pineapple Russet, d.—Above medium size, roundish ovate, and angular. Skin greenish yellow, dotted with white spots on one side, and covered with thick yellowish russet on the other. Stalk an inch long. Eye small, set in a shallow, plaited basin. Flesh pale yellow, crisp, very juicy, tender, with a highly aromatic perfume. One of the best dessert apples. September and October.

Pinner Seedling, d. (*Carel's Seedling*). — Medium sized, roundish ovate, and slightly angular. Skin greenish yellow, nearly covered with russet, and with a reddish-brown cheek next the sun. Eye small and closed. Stalk short. Flesh yellowish, crisp, juicy, sugary, and brisk. December to April.

PITMASTON NONPAREIL, D. (*Russet Coat Nonpareil*).—Medium sized, flat, compressed at the ends. Skin dull green, covered with a thin yellow russet, with a faint red next the sun. Stalk short. Eye open, large, and placed in a broad, shallow, and irregularly plaited cavity. Flesh greenish yellow, firm, and richly aromatic. November and December.

POMME GRISE, D.—Small, roundish, or ovate. Skin russety, with a brownish-red tinge on the side next the sun. Eye small and open. Stalk short. Flesh yellowish, tender, crisp, juicy, sugary, and aromatic. October to February.

Portugal. See *Reinette du Canada.*

Potter's Large. See *Kentish Fill Basket.*

POWELL'S RUSSET, D.—Small, roundish, flat at the ends. Skin yellowish green, and russety. Stalk short and slender. Eye small. Flesh pale yellow, rich, and aromatic. November to January.

Princess Noble. See *Golden Reinette.*

Princess Noble Zoete. See *Court-pendu Plat.*

Queen's Apple. See *Borsdörffer.*

RAVELSTON PIPPIN, D.—Medium sized, roundish, irregularly shaped, and ribbed. Skin greenish yellow, covered with red streaks. Eye closed. Stalk short and thick. Flesh yellow, firm, sweet, and pleasantly flavoured. August.

Read's Baker. See *Norfolk Beefing.*

RED ASTRACHAN, D. (*Anglesea Pippin*).—Medium sized, conical, and angular. Skin entirely covered with bright red on the side next the sun; deep yellow in the shade, and covered with a bloom. Stalk short, deeply inserted. Eye closed. Flesh white, and richly flavoured. Very prolific, and an excellent dessert apple. August and September.

Red Quarrenden. See *Devonshire Quarrenden.*

RED-STREAK, C.K. (*Scudamore's Crab*).—Medium sized, roundish. Skin deep yellow, and streaked all over with red. Eye small. Stalk short and slender. Flesh firm, crisp, and rather dry.

REINETTE BLANCHE D'ESPAGNE, K.D. (*Cobbett's Fall Pippin, Fall Pippin*).—Large, roundish, oblong, angular, with broad ribs; apex nearly as broad as the base. Skin yellowish green in the shade; orange, tinged with red, next the sun; dotted with black. Stalk half an inch long, set in a small cavity. Eye large, open, deeply sunk in an irregular basin. Flesh yellowish white, crisp, tender, and sugary. One of the largest apples, and of excellent quality. November to March.

Reinette du Canada, k.d. (*Portugal Russet, St. Helena Russet*).—Large, conical, and flattened. Skin greenish yellow, brown next the sun. Stalk short, inserted in a wide hollow. Eye set in a deep, irregular basin. Flesh white, firm, and juicy. November to May.

Reinette Grise, d.—Medium sized, round, and compressed at both ends, rather broadest at the base. Skin yellowish green in the shade; dull orange, tinged with red, next the sun; covered with grey russet. Eye small, set in a rather shallow, narrow, and angular basin. Stalk short, inserted in a wide and deep cavity. Flesh yellowish white, firm, juicy, rich, and sugary, with a pleasant subacid flavour. A dessert apple of the first quality. In use from November to April.

Reinette Van Mons, d.—Below medium size, flattened, and almost oblate, having five rather obscure ribs, which terminate in distinct ridges round the eye. Skin greenish yellow in the shade, but with a dull and brownish-orange tinge next the sun; the whole surface has a thin coating of brown russet. Eye closed, set in a rather deep depression. Stalk half an inch long. Flesh yellowish, tender, crisp, juicy, sugary, and aromatic. December to May.

Rhode Island Greening, k.d.—Large, roundish, and compressed. Skin dark green, changing to pale green, dullish red near the stalk, which is long, curved, and thickest at the bottom. Eye small, closed, and sunk in an open cavity. Flesh yellow, tender, crisp, juicy, rich, and aromatic. November to March.

Ribston Pippin, d. (*Glory of York, Formosa Pippin, Travers' Pippin*).—Medium sized, roundish, broadest at the base, irregular. Skin greenish yellow, changing when ripe to deep yellow; mottled and streaked with red and russet next the sun. Stalk half an inch long, slender, set in a wide cavity. Eye small, closed, and sunk in an irregular basin. Flesh deep yellow, fine, crisp, sharp, and richly aromatic. The king of English dessert apples. October to May.

Rook's Nest. See *Aromatic Russet.*

Rosemary Russet, d.—Medium sized, ovate. Skin yellow, tinged with green, tinged with red on the side next the sun, and covered with thin pale brown russet. Eye small and generally closed. Stalk very long. Flesh yellow, crisp, tender, very juicy, sugary, and highly aromatic. A first-rate dessert apple. In use from December to February.

Ross Nonpareil, d.—Medium sized, roundish, narrowing towards the eye. Skin covered with thin russet, with faint red next the sun. Stalk long, slender, and deeply inserted. Eye set in a shallow basin. Flesh greenish white, tender, and richly aromatic. November to April.

Round Winter Nonesuch, k.—Large, roundish, and depressed. Skin lively green, almost entirely covered with broken streaks and patches of fine deep red, and thickly strewed with russety dots; in some specimens the colour extends almost entirely round the fruit in long, broad patches. Eye large, closed, and prominetly set on the surface. Stalk short, deeply inserted in a funnel-shaped cavity. Flesh yellowish, firm, crisp, juicy, and slightly acid. A first-rate kitchen apple. In use from November to February.

Royal Pearmain, k.d. (*Herefordshire Pearmain*).—Rather large, oblong, and slightly angular. Skin yellowish green in the shade, and marked with russety specks; tinged with dull red next the sun, and sometimes with a few stripes of red. Eye small and open, set in a small, shallow basin. Stalk short, deeply inserted. Flesh yellowish, firm, crisp, juicy, and particularly rich and aromatic. In season from November to March.

Royal Russet, k. (*Leathercoat*).—Large, conical. Skin yellowish green, covered with grey russet. Stalk short. Eye small. Flesh greenish white, slightly aromatic. November to May.

Royal Somerset. See *London Pippin.*

Roxbury Russet. See *Boston Russet.*

Russetcoat Nonpareil. See *Pitmaston Nonpareil.*

Russet Golden Pippin. See *Golden Pippin.*

Russet Table Pearmain, d.—Below the medium size,

oblong ovate. Skin very russety, with yellowish green shining out on the shaded side; and orange, with a flame of red breaking through the russet, on the side next the sun. Eye open. Stalk half an inch long. Flesh yellow, firm, sugary, rich, and juicy, with a pleasant perfume. A first-rate dessert apple. In use from November to February.

Russian. See *Court-pendu Plat.*

Russian Emperor. See *Emperor Alexander.*

Rymer, k.—Large, roundish, regularly formed, and angular. Skin pale yellow, tinged all over with delicate rose, and of a deep bright red next the sun. Eye open, placed in an irregular, angular basin, which is surrounded by several prominent knobs. Stalk short and deeply inserted. Flesh yellow, delicate, juicy, and briskly acid. One of the best culinary apples, admirably adapted for sauce, or for baking. In use from October to December. Tree healthy, vigorous, and an abundant bearer.

Sack Apple. See *Devonshire Quarrenden.*

Sack and Sugar, d.—Below medium size, roundish, inclining to oval, with prominent ridges round the eye. Skin pale yellow, with a few broken streaks of red. Eye large and open. Stalk short. Flesh white, very tender, and juicy, with a brisk and balsamic flavour. Early in August.

St. Helena Russet. See *Reinette du Canada.*

Sam Rawlings. See *Hoary Morning.*

Sam Young, d. (*Irish Russet*).—Below medium size, roundish, compressed, and regular. Skin bright yellow, with grey russet, and dotted with brown spots on the yellow ground; russety red next the sun. Stalk short. Eye large and open, set in a broad basin. Flesh greenish yellow, tender, juicy, and richly flavoured. November to February.

Sam's Crab. See *Longville's Kernel.*

Scarlet Nonpareil, d.— Medium sized, roundish, flattened, handsome, and regularly formed. Skin green, tinged with russet; deep red, streaked with brown, next the sun. Stalk long and stout. Eye set in a regular, slightly-plaited, shallow basin. Flesh firm, yellowish white, rich, and juicy. January to March.

Scarlet Pearmain, d. (*Bell's Scarlet, Oxford Peach*).

—Medium sized, conical. Skin deep red, with yellow in the shade; bright crimson next the sun. Stalk long, slender, and deeply set. Eye full and deeply sunk. Flesh white, tinged with pink, juicy, crisp, and pleasant. September to December.

Screveton Golden Pippin, d.—Larger than the old Golden Pippin, and little, if at all, inferior to it in flavour. Skin yellowish, considerably marked with russet. Flesh yellow, and more tender than the old Golden Pippin. December to April.

Scudamore's Crab. See *Red-Streak.*

Shepherd's Seedling. See *Alfriston.*

Sir Walter Blackett's, d. (*Edinburgh Cluster*).—Small, roundish-ovate. Skin pale lemon yellow, very much dotted with pale brown russet and patches of the same, and with a faint orange tinge next the sun. Eye small and closed. Stalk long, deeply inserted. Flesh white, tender, juicy, and brisk, with a peculiar aroma. A first-rate dessert apple for northern districts. November to January.

Small's Admirable, k.d. — Above medium size, roundish-ovate, and flattened obtusely angular on the sides. Skin of a uniform lemon-yellow colour. Eye small, closed, and set in a rather deep basin. Stalk an inch long, slender. Flesh yellowish, firm, crisp, sweet, and agreeably acid, with a delicate perfume. November and December.

Somerset Lasting, k.—Large, oblate, and irregular on the sides. Skin pale yellow, streaked and dotted with a little bright crimson. Eye large and open. Stalk short. Flesh yellowish, tender, crisp, with a rough acid. October to February.

Sops in Wine, d.—Small, globular, narrow towards the eye. Skin crimson in the shade, stained and striped with purplish crimson next the sun; covered with white bloom. Stalk long and slender. Eye open, set in a shallow basin. Flesh white, stained with pink, firm, crisp, and juicy. October to February.

Speckled Golden Reinette. See *Barcelona Pearmain.*

Speckled Pearmain. See *Barcelona Pearmain.*

Spice Apple. See *Aromatic Russet.*

Springrove Codlin, k.—Medium sized, conical, broad at the base, and narrow at the apex. Skin greenish yellow, tinged with orange on the side next the sun. Eye closed. Stalk short. Flesh greenish yellow, soft, sweet, slightly acid, and agreeably perfumed. An excellent summer kitchen apple. In use from July to October.

Spring Ribston, d. (*Baddow Pippin*).—Below medium size, roundish, or rather oblate, ribbed on the sides, and knobbed round the apex. Skin yellowish green, covered with dull red next the sun. Eye large and open. Stalk very short. Flesh greenish white, crisp, juicy, sugary, with an aromatic flavour. A first-rate dessert apple. November to May.

Stagg's Nonpareil. See *Early Nonpareil.*

Stamford Pippin, d.k.—Large, roundish, inclining to ovate. Skin yellow, with a slight tinge of orange on one side. Eye small and half open. Stalk short. Flesh yellowish, firm, but quite tender, crisp, and very juicy, with a sweet, brisk flavour, and pleasant aroma. December to March.

Stettin Pippin. See *Dutch Mignonne.*

Striped Beefing, k.—Large, roundish, and somewhat flattened. Skin green, changing to greenish yellow, and almost entirely covered with broken streaks and patches of red. Eye large and open. Stalk rather short, deeply inserted. Flesh yellowish, firm, crisp, juicy, and pleasantly acid. A very fine culinary apple. In use from October to May.

Striped Holland Pippin. See *Lincolnshire Holland Pippin.*

Striped Joanneting. See *Margaret.*

Striped Quarrenden. See *Margaret.*

Sturmer Pippin, d.—Medium sized, roundish, and flattened. Skin yellowish green, and brownish red next the sun. Eye small and closed. Stalk long and straight. Flesh yellow, firm, brisk, sugary, and richly flavoured. January to June.

Sudlow's Fall. See *Franklin's Golden Pippin.*

Sugarloaf Pippin, k.—Medium sized, oblong. Skin fine light yellow, dotted with green; becoming almost white when fully ripe. Eye small. Stalk long. Flesh whitish, firm, crisp, and very juicy, with a pleasant, sweet, and sub-acid flavour. Ripe in August.

Summer Golden Pippin, d.—Below medium size,

ovate, and flattened at the ends. Skin pale yellow in the shade, tinged with orange and brownish red next the sun. Eye open. Stalk thick and short. Flesh yellowish, firm, very juicy, with a rich vinous and sugary flavour. A delicious, early dessert apple. End of August.

Summer Nonpareil. See *Early Nonpariel.*

SUMMER PEARMAIN, D. (*Autumn Pearmain*).—Medium sized, conical. Skin rich yellow, thickly dotted with brown dots in the shade, and striped and mottled with orange and bright red next the sun. Eye small. Stalk short, fleshy at the base, at its union with the fruit. Flesh yellowish white, crisp, and richly perfumed. October to January.

SUMMER STRAWBERRY, D.—Below medium size, oblate, even, and regularly formed. Skin smooth and shining, striped all over with yellow and red stripes. Eye not at all depressed, surrounded with prominent plaits. Stalk three quarters of an inch long. Flesh white, tender, juicy, briskly and pleasantly flavoured. September.

SWEENY NONPAREIL, K.—Above medium size, roundish-ovate. Skin bright green and russety, sometimes with a tinge of brown next the sun. Eye small and half open. Stalk long. Flesh greenish white, firm, crisp, and powerfully acid. An excellent sauce apple. January to April.

SYKEHOUSE RUSSET, D.—Small, roundish, compressed. Skin greenish yellow and russety in the shade, brownish red next the sun. Eye open, deeply sunk. Flesh greenish yellow, and richly flavoured. An excellent apple, and deserves general cultivation. November to February.

Thorle Pippin. See *Whorle Pippin.*

TOKER'S INCOMPARABLE, K.—Large, ovate, broad, and flattened at the base. Skin yellowish green, with a tinge of red next the sun, and a few crimson streaks. Eye large, nearly closed. Stalk very short. Flesh yellowish, firm, crisp, tender, juicy, with a pleasant acid. November to Christmas.

TOWER OF GLAMMIS, K. (*Carse of Gowrie*).—Large, conical, and distinctly four-sided. Skin deep sulphur yellow. Eye closed and deeply set. Stalk an inch long, deeply inserted. Flesh greenish white, very juicy, crisp, brisk, and perfumed. November to February.

Transparent Pippin. See *Court of Wick.*

Travers' Pippin. See *Ribston Pippin.*

TULIP, D.—Below medium size, ovato-conical. Skin, all over deep red, except on the shaded side, where it is golden yellow. Eye open. Stalk short. Flesh greenish yellow, crisp, juicy, sweet, and sub-acid. November to April.

WADHURST PIPPIN, K.—Above medium size, sometimes large, conical, and angular. Skin yellow, and mottled with brownish red on the side next the sun. Eye closed and deeply set. Stalk short and stout. Flesh yellowish, crisp, juicy, and briskly flavoured. October to February.

Walmer Court. See *Northern Greening.*

Waltham Abbey Seedling. See *Golden Noble.*

Warter's Golden Pippin. See *Golden Pippin.*

Warwickshire Pippin. See *Wyken Pippin.*

Week Pearmain. See *Wickham's Pearmain.*

Weeks' Pippin. See *Court of Wick.*

Wellington. See *Dumelow's Seedling.*

WHEELER'S RUSSET, D.—Medium sized, roundish, compressed, and irregular. Skin pale russet in the shade, bright brown next the sun. Stalk short. Eye small. Flesh greenish white, firm, and brisk flavoured. November to April.

WHITE ASTRACHAN, D.—Medium sized, conical. Skin pale yellow, or almost white, with faint streaks of red next the sun. Stalk thick and short. Eye small. Remarkable for the transparency of its flesh. August and September.

WHITE PARADISE, D. (*Lady's Finger, Egg Apple*).—Medium sized, oblong. Skin smooth, fine deep yellow, marked with broken streaks and dots of red. Eye open. Stalk an inch long. Flesh yellowish, tender, crisp, sugary, and pleasantly flavoured. October.

WHORLE PIPPIN, D. (*Thorle Pippin*).—Below medium size, oblate. Skin smooth, shining, and glossy, entirely covered with fine bright crimson, except where shaded, and then it is clear yellow. Eye large, half open, and frequently rent. Stalk very short. Flesh yellowish white, firm, crisp, very juicy, with a pleasant, refreshing flavour. August.

WICKHAM'S PEARMAIN, D. (*Week Pearmain*).—Medium sized, conical. Skin yellow, and almost entirely covered with bright red next the sun. Eye half open. Stalk half an inch long. Flesh greenish yellow, tender, crisp, juicy, sugary, and highly flavoured. October to December.

WINTER CODLIN, K.—Large, conical, five-sided, and ribbed. Skin smooth, yellowish green, and sometimes with a tinge of red next the sun. Eye very large and open. Stalk very short. Flesh greenish white, tender, juicy, sweet, and sub-acid. September to February.

WINTER COLMAN, K. (*Norfolk Colman, Norfolk Storing*).—Large, round, and much flattened at both ends. Skin pale yellow, spotted with red on the shaded side, and lively red next the sun. Eye small and open. Stalk short and deeply inserted. Flesh firm, crisp, and briskly acid. An excellent culinary apple. From November to March.

Winter Greening. See *French Crab*.

WINTER MAJETIN, K.—Large, roundish, terminated at the apex by five prominent crowns. Skin green, tinged with dull red on the side next the sun. Eye small and closed, set in a deep, narrow, and angular basin. Stalk long and slender. Flesh greenish white, firm, and of an agreeable acid flavour. This is a very desirable culinary apple. In season from November to March. The tree is a very prolific bearer.

WINTER PEARMAIN, K.D. (*Old Pearmain*).—Large, conical, somewhat five-sided towards the crown. Skin smooth and shining; greenish yellow on the shaded side, but covered with deep red and red streaks next the sun. Eye large and open. Stalk short. Flesh yellowish, firm, crisp, juicy, and sugary, with a brisk and pleasant flavour. December to April.

WINTER STRAWBERRY, K.D.—Medium sized, round. Skin yellow, striped with red. Eye prominent, surrounded with plaits. Stalk about an inch long, inserted in a shallow cavity. Flesh yellowish, crisp, juicy, briskly acid, and with a pleasant aroma. November to March.

WINTER QUOINING, K.D. (*Winter Queening*).—Medium sized, abrupt-conical, five-sided, and angular at the apex. Skin pale yellow, almost entirely covered with red. Eye small and closed. Stalk half an inch long. Flesh greenish yellow, tender, sweet, and perfumed. November to May.

Wollaton Pippin. *Court-pendu Plat.*

Wood's Huntingdon. See *Court of Wick.*

Woodstock Pippin. See *Blenheim Orange.*

WORMSLEY PIPPIN, K.D. (*Knight's Codlin*).—Medium sized, roundish, narrow towards the eye. Skin pale green, becoming deeper towards the sun, and marked with dark specks. Stalk an inch long, deeply set. Eye deeply sunk, placed in a plaited basin. Flesh white, crisp, and highly flavoured. September and October.

Wygers. See *Golden Reinette.*

WYKEN PIPPIN, D. (*Arley, Girkin Pippin, Warwickshire Pippin*).—Small, roundish, and compressed. Skin pale yellowish green, with dull orange next the sun. Stalk short. Eye small. Flesh greenish yellow, tender, very juicy, sweet, and richly flavoured. December to April.

Wyker Pippin. See *Golden Reinette.*

Yellow Harvest. See *Early Harvest.*

YORKSHIRE GREENING, K. (*Coates', Yorkshire Goose Sauce*).—Large, roundish, irregular, and flattened. Skin dark green, striped with dull red next the sun. Stalk short and thick. Eye closed. Flesh white, and pleasantly acid. One of the best kitchen apples. October to January.

---

## LISTS OF SELECT APPLES,

ADAPTED TO VARIOUS LATITUDES OF GREAT BRITAIN.

### I. SOUTHERN DISTRICTS OF ENGLAND,

AND NOT EXTENDING FURTHER NORTH THAN THE RIVER TRENT.

#### 1. SUMMER APPLES.

DESSERT.

Borovitsky
Devonshire Quarrenden
Early Harvest
Early Julien
Irish Peach
Joanneting
Kerry Pippin
Margaret
Sack and Sugar
Summer Golden Pippin

KITCHEN.

Carlisle Codlin
Duchess of Oldenburgh
Keswick Codlin
Lord Suffield
Manks Codlin
Springrove Codlin

## 2. Autumn Apples.

### Dessert.

Adams' Pearmain
American Mother Apple
Borsdörffer
Blenheim Pippin
Claygate Pearmain
Coe's Golden Drop
Cornish Aromatic
Court of Wick
Cox's Orange Pippin
Downton Pippin
Early Nonpareil
Fearn's Pippin
Franklin's Golden Pippin
Golden Pippin
Golden Reinette
Golden Winter Pearmain
Lucombe's Pine
Margil
Melon Apple
Nanny
Pine Apple Russet
Ribston Pippin
Sykehouse Russet
Red Ingestrie
Reinette Van Mons
White Ingestrie

### Kitchen.

Bedfordshire Foundling
Cellini
Cox's Pomona
Emperor Alexander
Flower of Kent
Forge
Gloria Mundi
Golden Noble
Greenup's Pippin
Harvey Apple
Hawthornden
Hoary Morning
Kentish Fill Basket
Lemon Pippin
Mère de Ménage
Nelson Codlin
Nonesuch
Tower of Glammis
Wadhurst Pippin
Winter Quoining
Wormsley Pippin
Yorkshire Greening

## 3. Winter Apples.

### Dessert.

Ashmead's Kernel
Barcelona Pearmain
Boston Russet
Braddick's Nonpareil
Claygate Pearmain
Cockle Pippin
Cornish Gilliflower
Court-pendu Plat
Downton Nonpareil
Dredge's Fame
Dutch Mignonne
Golden Harvey
Golden Russet
Hughes' Golden Pippin
Hubbard's Pearmain
Keddleston Pippin
Lamb Abbey Pearmain
Maclean's Favourite
Mannington's Pearmain
Nonpareil
Ord's Apple
Pearson's Plate
Pinner Seedling
Pitmaston Nonpareil
Ross Nonpareil
Russet Table Pearmain
Sam Young
Spring Ribston
Sturmer Pippin
Wyken Pippin

### Kitchen.

Alfriston
Beauty of Kent
Bess Pool

Brabant Bellefleur
Brownlees' Russet
Dumelow's Seedling
French Crab
Gooseberry Apple
Hambledon Deux Ans
Hanwell Souring
Minchall Crab
Norfolk Beefing
Norfolk Colman
Norfolk Stone Pippin
Northern Greening
Reinette Blanche d'Espagne
Rhode Island Greening
Round Winter Noncsuch
Royal Pcarmain
Royal Russet
Striped Beefing
Winter Majetin
Winter Pearmain

---

## II. NORTHERN DISTRICTS OF ENGLAND,

EXTENDING FROM THE RIVER TRENT TO THE RIVER TYNE.

### 1. Summer Apples.

Dessert.

Devonshire Quarrenden
Early Harvest
Irish Peach
Joanneting
Kerry Pippin
Margaret
Oslin
Whorle

Kitchen.

Carlisle Codlin
Keswick Codlin
Lord Suffield
Manks Codlin
Nonesuch
Springrove Codlin

### 2. Autumn Apples.

Dessert.

Borsdörffer
Downton Pippin
Early Nonpareil
Franklin's Golden Pippin
Golden Monday
Golden Winter Pearmain
Red Ingestrie
Ribston Pippin
Stamford Pippin
Summer Pearmain
Wormsley Pippin
Yellow Ingestrie

Kitchen.

Cellini
Emperor Alexander
Gloria Mundi
Greenup's Pippin
Hawthornden
Lemon Pippin
Mère de Ménage
Nelson Codlin
Nonesuch
Tower of Glammis

### 3. Winter Apples.

Dessert.

Adams' Pearmain
Barcelona Pearmain
Bess Pool
Braddick's Nonpareil
Claygate Pearmain
Cockle Pippin
Court of Wick
Court-pendu Plat
Golden Pippin

**Dessert.**

Golden Reinette
Keddleston Pippin
Margil
Nonpareil
Pitmaston Nonpareil
Royal Pearmain
Scarlet Nonpareil
Sturmer Pippin
Sykehouse Russet

**Kitchen.**

Alfriston
Bedfordshire Foundling
Blenheim Pippin
Dumelow's Seedling
French Crab
Mère de Ménage
Nelson Codlin
Northern Greening
Round Winter Nonesuch
Yorkshire Greening

---

## III. BORDER COUNTIES of ENGLAND and SCOTLAND, AND THE WARM AND SHELTERED SITUATIONS IN OTHER PARTS OF SCOTLAND.

### 1. Summer and Autumn Apples.

**Dessert.**

Bess Pool
Cellini
Devonshire Quarrenden
Early Julien
Federal Pearmain
Golden Monday
Greenup's Pippin
Grey Leadington
Irish Peach
Kerry Pippin
Margaret
Nonesuch
Oslin
Ravelston Pippin
Red Astrachan
Red Ingestrie
Sir Walter Blackett's
Summer Pearmain
Summer Strawberry
White Paradise
Whorle
Wormsley Pippin
Yellow Ingestrie

**Kitchen.**

Carlisle Codlin
Dutch Codlin
Hawthornden
Keswick Codlin
Manks Codlin
Nelson Codlin
Springrove Codlin

### 2. Winter Apples.

*Those marked * require a wall.*

**Dessert.**

*Adams' Pearmain
*Barcelona Pearmain
Bess Pool
*Braddick's Nonpareil
Court of Wick
*Downton Pippin
*Golden Pippin
*Golden Russet
*Margil
*Nonpareil
*Pearson's Plate
*Pennington's Seedling
*Ribston Pippin

Scarlet Nonpareil
Sturmer Pippin
Sykehouse Russet
Wyken Pippin

KITCHEN.

Bedfordshire Foundling
Brabant Bellefleur
Dumelow's Seedling
French Crab
Royal Russet
Rymer
Tower of Glammis
Winter Pearmain
Winter Strawberry
Yorkshire Greening

---

## IV. NORTHERN PARTS OF SCOTLAND,

### AND OTHER EXPOSED SITUATIONS IN ENGLAND AND SCOTLAND.

### 1. SUMMER AND AUTUMN APPLES.

*Those marked * require a wall.*

DESSERT.

Devonshire Quarrenden
Early Julien
Kerry Pippin
Nonesuch
*Ravelston Pippin
Summer Strawberry

KITCHEN.

Carlisle Codlin
Hawthornden
Keswick Codlin
Manks Codlin

### 2. WINTER APPLES.

DESSERT.

*Golden Russet
Grey Leadington
*Margil
Winter Strawberry

KITCHEN.

French Crab
Tower of Glammis
Yorkshire Greening

---

## V. FOR ESPALIERS, OR DWARFS.

These succeed well when grafted on the Paradise or Doucin stock; and, from their small habit of growth, are well adapted for that mode of culture.

Adams' Pearmain
American Mother Apple
Ashmead's Kernel
Borovitski
Boston Russet
Braddick's Nonpareil
Cellini
Christie's Pippin
Claygate Pearmain
Cockle Pippin
Coe's Golden Drop
Cornish Gilliflower
Court of Wick
Court-pendu Plat
Cox's Orange Pippin
Downton Pippin

Dutch Mignonne
Early Harvest
Early Julien
Early Nonpareil
Franklin's Golden Pippin
Golden Harvey
Golden Pippin
Golden Reinette
Golden Russet
Hawthornden
Holbert's Victoria
Hubbard's Pearmain
Hughes' Golden Pippin
Irish Peach
Isle of Wight Pippin
Joanneting
Keddleston Pippin
Kerry Pippin
Keswick Codlin
Lamb Abbey Pearmain
Lucombe's Pine
Maclean's Favourite
Manks Codlin
Mannington's Pearmain
Margaret
Margil
Melon Apple
Nanny
Nonesuch
Nonpareil
Oslin
Pearson's Plate
Pennington's Seedling
Pine Apple Russet
Pinner Seedling
Pitmaston Nonpareil
Red Ingestrie
Reinette Van Mons
Ross Nonpareil
Russet Table Pearmain
Sam Young
Scarlet Nonpareil
Scarlet Pearmain
Sturmer Pippin
Summer Golden Pippin
Summer Pearmain
Sykehouse Russet
Yellow Ingestrie

---

## VI. FOR ORCHARD PLANTING AS STANDARDS.

These are generally strong-growing or productive varieties, the fruit of which being mostly of a large size, or showy appearance, they are, on that account, well adapted for orchard planting, to supply the markets.

Alfriston
Barcelona Pearmain
Beauty of Kent
Bedfordshire Foundling
Bess Pool
Blenheim Pippin
Brabant Bellefleur
Brownlees' Russet
Cellini
Cox's Pomona
Devonshire Quarrenden
Duchess of Oldenburgh
Dumelow's Seedling
Dutch Codlin
Emperor Alexander
English Codlin
Fearn's Pippin
Flower of Kent
Forge
French Crab
Gloria Mundi
Golden Noble
Golden Winter Pearmain
Gooseberry Apple
Hambledon Deux Ans
Hanwell Souring
Harvey Apple
Hoary Morning

Hollandbury
Kentish Fill Basket
Kerry Pippin
Keswick Codlin
Lemon Pippin
Lewis' Incomparable
London Pippin
Longvilles' Kernel
Manks Codlin
Margaret
Mère de Ménage
Minchall Crab
Minier's Dumpling
Nelson Codlin
Norfolk Bearer
Norfolk Beefing
Northern Greening
Reinette Blanche d'Espagne
Reinette du Canada
Rhode Island Greening
Round Winter Nonesuch
Royal Pearmain
Royal Russet
Rymer
Small's Admirable
Striped Beefing
Toker's Incomparable
Tower of Glammis
Winter Codlin
Winter Colman
Winter Majetin
Winter Pearmain
Winter Quoining
Wormsley Pippin
Wyken Pippin
Yorkshire Greening

# APRICOTS.

## SYNOPSIS OF APRICOTS.

### I. KERNELS BITTER.

#### * *Back of the stone impervious.*

A. *Freestones.*

Brussels
Large Early
Pine Apple
Red Masculine
Roman
Royal
St. Ambroise
Shipley's
White Masculine

B. *Clingstones.*

Black
Montgamet
Portugal

#### ** *Back of the stone pervious* (1).

Alsace
Hemskerk
Large Red
Moorpark
Peach
Viard

### II. KERNELS SWEET.

A. *Freestones.*

Angoumois
Breda
Kaisha
Musch Musch
Turkey
Provence

B. *Clingstones.*

Orange

---

Alberge de Montgamet. See *Montgamet.*

D'Alexandrie. See *Musch Musch.*

ALSACE.—This is a variety of the Moorpark, and is of a very large size, with a rich and juicy flavour; and the tree, unlike the others of the race, is vigorous and hardy, and does not die off in branches, as the Moorpark does.

Amande Aveline. See *Breda.*

(1) The bony substance at the back of the stone is pervious by passage, through which a pin may be passed from one end to the other.

Angoumois (*Violet; Anjou; Rouge*).—Small, oval, flattened at the apex, and marked on one side with a shallow suture, the sides of which are raised. Skin clear, deep yellow on the shaded side, but dark rusty brown on the side next the sun. Flesh deep orange, juicy, and melting, separating from the stone; rich, sugary, and briskly flavoured; but, when highly ripened, charged with a fine aroma. Back of the stone impervious. Kernel sweet. End of July.

Ananas. See *Pine Apple.*

Anjou. See *Angoumois.*

Anson's. See *Moorpark.*

Aveline. See *Breda.*

Black (*Noir; Purple*).—About the size and shape of a small Orleans plum, to which it bears some resemblance. Skin of a deep black-purple colour next the sun, but paler on the shaded side, and covered with delicate down. Flesh pale red, but darker near the stone; juicy, but tasteless and insipid, and quite worthless to eat. Stone small, impervious on the back. Kernel bitter. Ripe in the beginning of August.

Blanc. See *White Masculine.*

Blenheim. See *Shipley's.*

Breda (*Aveline; Amande Aveline*).—Rather small, roundish, compressed on the sides, and sometimes entirely four-sided. Skin deep orange, dotted with brown spots next the sun. Suture well defined. Flesh deep orange, rich, highly flavoured, and free. Stone small, roundish, impervious on the back. Kernel sweet, with the flavour of a hazel-nut. End of August.

Brussels.—Medium sized, rather oval, flattened on the sides. Skin pale yellow, dotted with white; red, interspersed with dark spots, next the sun. Suture deep next the stalk, diminishing towards the apex. Flesh yellow, firm, brisk flavoured, and free. Stone small, impervious on the back. Kernel bitter. The best to cultivate as a standard. Middle of August.

Common. See *Roman.*

Crotté. See *Montgamet.*

Dunmore's. See *Moorpark.*

Early Orange. See *Portugal.*

Gros d'Alexandrie. See *Large Early.*

Gros Commun. See *Roman.*

Gros Pêche. See *Peach.*

Gros Précoce. See *Large Early.*

Gros Rouge. See *Large Red.*

Hemskerk.—Rather large, round, flattened on the sides. Skin orange, reddish next the sun. Suture distinct, higher on one side than the other. Flesh bright orange, tender, rich, and juicy, separating from the stone. Stone small, pervious on the back. Kernel bitter. This very much resembles, and, according to some, equals, the Moorpark. The tree is certainly hardier than that variety. End of July and beginning of August.

Hunt's Moorpark. See *Moorpark.*

Kaisha.—Medium sized, roundish, marked with a suture, which is deep towards the stalk, and gradually diminishes towards the apex, which is pitted. Skin pale-lemon coloured on the shaded side, and tinged and mottled with red next the sun. Flesh transparent, separating freely from the stone, clear pale yellow, tender, and very juicy, sugary, and richly flavoured. Stone small, roundish. Kernel sweet. Middle of August.

Large Early (*Précoce d'Esperen; Gros Précoce; Gros d'Alexandrie; De St. Jean; Précoce d'Hongrie*).—Above the medium size, rather oblong, and flattened on the sides. Skin pale orange on the shaded side; bright orange, and spotted with red, next the sun; slightly downy. Suture deep. Flesh deep orange, rich, juicy, separating from the stone, which is very flat, oval, sharp at the point, and impervious on the back. Kernel bitter. End of July and beginning of August.

Large Red (*Gros Rouge*).—This is a variety of the Peach apricot, and of a deeper colour than that variety. It is large, and of a deep orange-red colour. The flesh is rich and juicy, and separates freely from the stone. Stone pervious along the back. Kernel bitter. The tree is said, by Mr. Rivers, who introduced this variety, to be hardier than the Moorpark.

Montgamet (*Crotté; Alberge de Montgamet*).—Of small size, oval, somewhat compressed on the sides, and marked with a shallow suture. Skin pale yellow, with a slight tinge of red on the side next the sun. Flesh yellowish, firm, adhering to the stone, juicy, and agreeably acid; but when well ripened it is highly perfumed. Stone impervious, roundish. Kernel bitter. Ripe in the end of July; and generally used for preserving.

Moorpark (*Anson's; Dunmore's; Hunt's Moorpark; Oldaker's Moorpark; Sudlow's Moorpark; Temple's*).—Large, roundish, more swollen on one side of the suture than the other. Skin pale yellow on the shaded side, and deep orange, or brownish red, next the sun, and marked with dark specks. Flesh bright orange, firm, juicy, and of rich luscious flavour; separating from the stone, which is rough and pervious on the back. Kernel bitter. End of August and beginning of September.

Musch Musch (*D'Alexandrie*).—Small, almost round, and slightly compressed. Skin deep yellow; orange red next the sun. Flesh yellow, remarkably transparent, tender, melting, and the sweetest of all apricots. Stone impervious. Kernel sweet. Excellent for preserving. Ripe in the end of July.

De Nancy. See *Peach*.

Noir. See *Black*.

Oldaker's Moorpark. See *Moorpark*.

Peach (*Pêche; Gros Pêche; De Nancy; De Wirtemberg; Royal Peach*).—Large, oval, and flattened, marked with a deep suture at the base, which gradually diminishes towards the apex. Skin pale yellow on the shaded side, and a slight tinge of red next the sun. Flesh reddish yellow, very delicate, juicy, and sugary, with a rich and somewhat musky flavour. Stone large, flat, rugged, and pervious along the back. Kernel bitter. This is quite distinct from the Moorpark, now cultivated under that name; and is, doubtless, the parent of all the varieties so called. It may always be distinguished from the Moorpark by nurserymen; for, while the Moorpark may be budded freely on the common plum stock, the Peach apricot will only take on the Muscle stock. Ripe the end of August and beginning of September.

Pêche. See *Peach*.

Pine Apple (*Ananas*).—Large, roundish, and flattened, and marked with a rather shallow suture. Skin thin and delicate, of a deep golden yellow on the shaded side, but with a highly-coloured red cheek where exposed to the sun, and speckled with large and small red specks. The flesh is reddish yellow, tender, but somewhat firm; never becomes mealy, but is juicy, and with a rich pine-apple flavour. Stone oval and three-ribbed, and impervious along the back. Kernel bitter. Ripens in the middle of August.

Portugal (*Early Orange*).—Very small, resembling, in shape and size, the Red Masculine. It is round, and divided on one side by a deep suture. Skin pale yellow on the shaded side, and deep yellow, tinged with red, and marked with brown and red russet spots on the side next the sun. Flesh deep yellow, tender, melting, with a rich sugary and musky flavour; adhering somewhat to the stone. Stone almost round, impervious along the back. Kernel bitter. Ripe in the beginning and middle of August.

Précoce d'Esperen. See *Large Early*.

Précoce d'Hongrie. See *Large Early*.

Purple. See *Black*.

Red Masculine.—Small, roundish. Skin bright yellow on the shaded side; deep orange, spotted with dark red, next the sun. Suture well defined. Flesh yellow, juicy, and musky. Stone thick, obtuse at the ends, impervious along the back. Kernel bitter. July.

Roman (*Common*).—Above medium size, oblong, sides compressed. Skin pale yellow, with rarely a few red spots next the sun. Suture scarcely perceptible. Flesh dull yellow, soft, and dry, separating from the stone, and possessing a sweet and agreeable acid juice, that makes it desirable for preserving. Stone oblong, impervious. Kernel bitter. Middle of August.

Rouge. See *Angoumois*.

Royal.—Large, oval, and slightly compressed. Skin dull yellow, tinged with red where exposed. Suture shallow. Flesh pale orange, firm, juicy, rich, and vinous, separating from the stone. Stone large and oval, impervious. Kernel bitter. An excellent apricot, and little inferior to the Moorpark. Beginning of August.

Royal Orange.—Above medium size, roundish, one side swelling more than the other. Skin pale orange in the shade; deep orange, tinged with red, next the sun. Suture well defined, deep towards the stalk. Flesh deep orange, firm, and adhering to the stone, which is small, smooth, thick, and impervious. Kernel sweet. Middle of August.

Royal Peach. See *Peach.*

St. Ambroise.—This is a large, early apricot, almost the size of the Moorpark. It is compressed, of a deep yellow colour, reddish next the sun. Flesh juicy, rich, and sugary. Ripe the middle of August. The tree is said to be very vigorous, healthy, and a good bearer.

De St. Jean. See *Large Early.*

Shipley's (*Blenheim*).—Large, oval. Skin deep yellow. Flesh yellow, tolerably rich and juicy. Stone roundish, and impervious. Kernel bitter. Very productive and early, but not so rich as the Moorpark. End of July and beginning of August.

Sudlow's Moorpark See *Moorpark.*

Tardive d'Orleans.—This is a late variety, ripening a fortnight after the Moorpark.

Temple's See *Moorpark.*

Turkey.—Medium size, nearly round, not compressed. Skin deep yellow; brownish orange next the sun, and spotted. Flesh pale yellow, firm, juicy, sweet, and pleasantly sub-acid, separating from the stone. Stone large, rugged, and impervious. Kernel sweet. Middle of August.

Viard.—This, according to Mr. Rivers, is an early variety of the Peach apricot, with rich, juicy flesh. The tree is hardy.

Violet. See *Angoumois.*

White Masculine (*Blanche*).—Small, round, and somewhat compressed at both ends. Skin covered with a fine white down; pale yellow, tinged with brownish red, next the sun, and dull white in the shade. Flesh pale yellow, adhering in some degree to the stone; fine and delicate, juicy, sugary, and excellent. Kernel bitter. Ripe the end of July.

De Wirtemberg. See *Peach.*

# LIST OF SELECT APRICOTS.

## I. FOR THE SOUTHERN COUNTIES OF ENGLAND

EXTENDING AS FAR NORTH AS THE RIVER TRENT.

*For Walls.*

Hemskerk
Kaisha
Large Early
Large Red
Moorpark
Peach
Pine Apple
Royal
Shipley's
Turkey

*For Standards.*

Breda
Brussels
Moorpark
Turkey

## II. FOR THE NORTHERN COUNTIES OF ENGLAND,

EXTENDING FROM THE TRENT TO THE TYNE.

Breda
Brussels
Hemskerk
Moorpark
Red Masculine
Roman
Royal Orange
Shipley's

## III. BORDER COUNTIES OF ENGLAND AND SCOTLAND,

AND OTHER FAVOURABLE SITUATIONS IN SCOTLAND.

Breda
Brussels
Hemskerk
Red Masculine
Roman
Royal Orange

## IV. VARIETIES BEST ADAPTED FOR PRESERVING.

Kaisha
Moorpark
Musch Musch
Peach
Roman
Turkey

## BERBERRIES.

Berberries, though not cultivated to any extent, ma be enumerated among the British fruits. The Common Berberry is found wild in hedgerows, and is also sometimes grown in shrubberies, both as an ornamental plant, and for its fruit, which is preserved in sugar, for use in the dessert. The best variety to cultivate for that purpose is the following, but it is difficult to be obtained true.

Stoneless Berberry.—A variety of the Common Berberry, without seeds. This character is not assumed till the shrub has become aged; for young suckers, taken from an old plant of the true variety, very frequently, and indeed generally, produce fertile fruit during the early years of their growth; it is, therefore, necessary to be assured that the plants were taken from an aged stock, in which the stoneless character had been manifested, to be certain that the variety is correct.

# CHERRIES.

## SYNOPSIS OF CHERRIES.

In the following arrangement I have endeavoured to classify all those varieties of cherries that are most nearly allied to each other, for the purpose of facilitating their identification.

All the varieties of cultivated cherries will be found to consist of eight races, into which I have arranged them:—I. The sweet, heart-shaped cherries, with tender and dark-coloured flesh, I have called Black-Geans. II. The pale-coloured, sweet cherries, with tender, light yellow, and translucent flesh and skin, I have distinguished by the name of Amber-Geans, as at once expressive of their character. III. Here we have the dark-coloured, sweet cherries, with somewhat of the Bigarreau character. Their flesh is not so firm and crackling as that of the Bigarreaus, but considerably harder than in the Black Geans, and these I propose to call Hearts. IV. Includes the Bigarreaus, properly so called, with light-coloured mottled skin, and hard, crackling flesh. V. These are called Dukes, as they include all those so well known under that name. VI. Embraces all those nearly allied to the Dukes, but with pale-red skin, translucent skin and flesh, and uncoloured juice; they are, therefore, distinguished as Red-Dukes. VII. Includes all those, the trees of which have long, slender, and pendent shoots, and dark-coloured fruit, with acid, coloured juice, and appropriately termed Morellos; and VIII. I have called Kentish, as it includes all those pale-red, acid varieties, of which the Kentish cherry is the type.

The advantages of such an arrangement and nomenclature are, that they not only facilitate identification, but assist description and interchange of ideas. If, for instance, a new cherry is introduced, and it is said to belong to the Red-Dukes, we know at once that it has some affinity with those familiar varieties Belle de Choisy and Carnation; or if it be a Morello, we know it is a dark-fleshed, acid cherry; while if we are told it is a Kentish, then we know it is a pale-fleshed, acid variety; and so with all the other divisions.

### I. GEANS.

Branches rigid and spreading, forming round-headed trees. Leaves long, waved on the margin, thin and flaccid, and feebly supported on the footstalks. Flowers large, and opening loosely, with thin, flimsy, obovate, or roundish-ovate, petals. Fruit heart-shaped, or nearly so. Juice sweet.

§ *Fruit obtuse heart-shaped. Flesh tender and melting.*

* *Flesh dark; juice coloured.*—BLACK-GEANS.

Baumann's May
Black Eagle
Early Purple Gean
Hogg's Black Gean
Joc-o-sot
Knight's Early Black
Luke Ward's
Osceola
Waterloo
Werder's Early Black

** *Flesh pale; juice uncoloured.*—AMBER-GEANS.

Amber Gean
American Doctor
Belle d'Orléans
Delicate
Downer's Late
Early Amber
Hogg's Red Gean
Manning's Mottled
Ohio Beauty
Sparhawk's Honey
Transparent Gean

§§ *Fruit heart-shaped. Flesh half-tender, firm, or crackling.*

* *Flesh dark; juice coloured.*—HEARTS.

Black Hawk
Black Heart
Black Tartarian
Brant
Büttner's Black Heart
Corone
Logan
Monstrous Heart
Ox-Heart
Pontiac
Powhattan
Tecumseh
Tradescant's Heart

** *Flesh pale; juice uncoloured.*—BIGARREAUS.

Adams' Crown
Belle Agathe
Bigarreau
Bigarreau de Hildesheim
Bigarreau de Hollande
Bigarreau Napoléon
Bowyer's Early Heart
Büttner's Yellow
Cleveland Bigarreau
Downton
Gascoigne's Heart
Harrison's Heart
Early Prolific
Elton
Florence
Governor Wood
Kennicott
Lady Southampton's
Late Bigarreau
Mammoth
Mary
Red Jacket
Rockport Bigarreau
Tardive de Mans
Tobacco-Leaved
White Heart
White Tartarian

II. GRIOTTES.

Branches either upright, spreading, or more or less long, slender, and drooping. Leaves flat, dark green, glabrous underneath, and borne stiffly on the leafstalks; large and broad in §, and small and narrow in §§. Flowers in pedunculate umbels, cup-shaped, with firm, stiff, and crumpled orbicular petals. Fruit round or oblate, sometimes, as in the Morello, inclining to heart-shaped. Juice sub-acid or acid.

§ *Branches upright, occasionally spreading. Leaves large and broad.*

* *Flesh dark; juice coloured.*—DUKES.

Archduke
Büttner's October
Duchesse de Palluau
Griotte de Chaux
Jeffreys' Duke
May Duke
Royal Duke
De Soissons

** *Flesh pale; juice uncoloured.*—RED-DUKES.

Belle de Choisy
Belle Magnifique
Carnation
Coe's Late Carnation
Great Cornelian
Reine Hortense
Late Duke

§§ *Branches long, slender, and drooping. Leaves small and narrow.*

* *Flesh dark; juice coloured.*—MORELLOS.

Double Natte
Early May
Griotte de Kleparow
Morello
Ostheim
Ratafia
Shannon Morello

** *Flesh pale; juice uncoloured.*—KENTISH.

All Saints
Cluster
Flemish
Gros Gobet
Kentish
Paramdam

---

ADAMS' CROWN.—Medium sized, obtuse heart-shaped, and slightly compressed on one side. Skin pale red, mottled with yellow. Stalk two inches long. Flesh white, tender, juicy, and richly flavoured. An excellent bearer, and a first-rate early cherry. Beginning of July.

ALL SAINTS' (*Autumn-bearing Cluster; De St. Martin; Toussaint*).—Small and oblate. Skin red, becoming dark red as it hangs on the tree. Stalk two inches long. Flesh white, reddish next the stone, juicy, and acid. This is generally grown as an ornamental tree.

Amarelle du Nord. See *Ratafia.*

AMBER GEAN.—Below medium size, generally in triplets; obtuse heart-shaped. Skin thin and transparent, pale yellow, or amber, tinged with delicate red. Stalk slender, an inch and a half long. Flesh white, tender, and juicy, with a rich, sweet, and delicious flavour. Beginning of August.

Amber Heart. See *White Heart.*

Ambrée. See *Belle de Choisy.*

AMERICAN AMBER (*Bloodgood's Amber; Bloodgood's Honey*).—Medium sized, growing in clusters; roundish, inclining to heart-shaped. Skin thin and shining, clear yellow, mottled with bright red. Stalk an inch and a half long. Flesh amber coloured. Beginning of July.

AMERICAN DOCTOR (*The Doctor*).—Medium sized, obtuse heart-shaped. Skin clear yellow, washed with red. Stalk an inch and a half long. Flesh yellowish white, tender, juicy, sweet, and richly flavoured. End of June. I have named this "American Doctor" to distinguish it from the German "Doktorkirsche."

Anglaise Tardive. See *Late Duke.*

Angleterre Hâtive. See *May Duke.*

Ansell's Fine Black. See *Black Heart.*

ARCHDUKE.—Larger than May Duke, obtuse heart-shaped, with a deeply-marked suture at the apex, diminishing towards the stalk, and very slightly pitted at the apex. Skin thin, pale red at first, but becoming dark red, and ultimately almost black. Stalk very slender, an inch and a half to two inches long. Flesh deep red, very tender and juicy, sweet, and briskly flavoured; but sugary when highly ripened. Middle and end of July. Tree somewhat pendulous when old.

D'Aremberg. See *Reine Hortense.*

Armstrong's Bigarreau. See *Bigarreau ae Hollande.*

Baramdam. See *Paramdam.*

BAUMANN'S MAY (*Bigarreau de Mai; Trempée Precoce*).—Medium sized, ovate, inclining to cordate, and irregular in its shape. Skin of a fine dark-red colour, changing to deep, shining black. Stalk about two inches long. Flesh purple, tender, juicy, and excellent. Ripe the middle of June.

Belcher's Black. See *Corone.*

BELLE AGATHE.—Small, produced in clusters; heart-shaped. Skin dark crimson, with minute yellow mottles over it. Stalk an inch and a half to an inch and three quarters long. Flesh yellowish, firm, sweet, and very nicely flavoured. This is a small Bigarreau, which hangs on the tree as late as the first week in October; and neither birds nor wasps touch it.

Belle Audigeoise. See *Reine Hortense.*

Belle de Bavay. See *Reine Hortense.*

Belle de Chatenay. See *Belle Magnifique.*

BELLE DE CHOISY (*Ambrée; Dauphine; Doucette; De Palembre*).—Large and round. Skin very thin and transparent, showing the texture of the flesh beneath; amber coloured, mottled with yellowish red, or rich cornelian, next the sun. Stalk an inch and a half to two inches long, rather stout, swollen at the upper end. Flesh amber coloured, melting, tender, rich, sugary and delicious. Early in July.

Belle de Laeken. See *Reine Hortense.*

BELLE MAGNIFIQUE (*Belle de Chatenay; Belle de Sceaux; Belle de Spa; De Spa*).—Very large, roundish-oblate, inclining to heart-shaped. Skin clear bright red. Stalk an inch to an inch and a half long. Flesh yellowish, tender, and sub-acid. Middle and end of August.

BELLE D'ORLEANS.—Medium sized, roundish, inclining to heart-shaped. Skin pale yellowish white in the shade, but of a thin bright red next the sun. Flesh yellowish white, tender, juicy, and rich. Beginning and middle of June. One of the earliest and richest of cherries.

Belle de Petit Brie. See *Reine Hortense.*

Belle Polonaise. See *Griotte de Kleparow*

Belle de Prapeau. See *Reine Hortense.*

Belle de Sceaux. See *Belle Magnifique.*

Belle de Spa. See *Belle Magnifique.*

Belle Suprême. See *Reine Hortense.*

BIGARREAU (*Graffion*).—Large, and obtuse heart-shaped, flattened at the stalk. Skin whitish yellow, marbled with deep bright red next the sun. Stalk stout, two inches long, deeply inserted. Flesh pale yellow, firm, rich, and highly flavoured. Stone large and round. End of July.

Bigarreau Gaboulais. See *Monstrous Heart.*

Bigarreau Gros Cœuret. See *Monstrous Heart.*

Bigarreau Gros Monstrueux. See *Monstrous Heart.*

Bigarreau Gros Noir. See *Tradescant's Heart.*

BIGARREAU DE HILDESHEIM (*Bigarreau Tardif de Hildesheim*).—Medium sized, heart-shaped, flattened on one side. Skin shining, pale yellow, marbled with red on one side, but dark red on the other. Stalk two inches long. Flesh yellow, very firm, not particularly juicy, but with an excellent sweet flavour. Ripe the end of August and beginning of September. An excellent late cherry.

BIGARREAU DE HOLLANDE (*Spotted Bigarreau; Armstrong's Bigarreau*).—Very large, regularly and handsomely heart-shaped. Skin pale yellow on the shaded side, but of a light red, marbled with bright crimson, on the side exposed to the sun. Stalk an inch and a half long, stout, inserted a little on one side of the fruit. Flesh pale yellowish white, juicy, and sweet, with an agreeable piquancy. Stone small for the size of the fruit. Middle of July.

Bigarreau Jaboulais. See *Monstrous Heart.*

Bigarreau Lauermann. See *Bigarreau Napoléon.*

Bigarreau de Lyons. See *Monstrous Heart.*

Bigarreau de Mai. See *Baumann's May.*

Bigarreau Monstrueux de Mezel. See *Monstrous Heart.*

Bigarreau Monstrueux. See *Monstrous Heart.*

BIGARREAU NAPOLEON (*Bigarreau Lauermann*).—Large, and oblong heart-shaped. Skin pale yellow, spotted with deep red, marbled with fine deep crimson next the sun. Stalk stout and short, set in a narrow cavity. Flesh very firm, juicy, and of excellent flavour. An abundant bearer. July and August.

Bigarreau Tardif de Hildesheim. See *Bigarreau de Hildesheim.*

Black Bud of Buckinghamshire. See *Corone.*

Black Caroon. See *Corone.*

Black Circassian. See *Black Tartarian.*

Black Eagle.—Medium sized, obtuse heart-shaped, compressed at both ends. Skin deep purple; when ripe nearly black. Stalk an inch and a half long, and slender. Flesh deep purple, tender, very rich, and juicy. Beginning of July.

Black Hawk.—Large, obtuse heart-shaped, uneven in its outline, and compressed on the sides. Skin deep, shining, blackish-purple. Stalk about an inch and a half long. Flesh dark purple, tolerably firm, rich, and sweet. Middle and end of July.

Black Heart (*Ansell's Fine Black; Early Black; Lacure; Spanish Black Heart*).—Above medium size, heart-shaped, rather irregular, compressed at the apex. Skin dark purple; deep black when quite ripe. Stalk an inch and a half long, slender. Flesh half tender, rich, juicy, and sweet. Early in July.

Black Morello. See *Morello.*

Black Orleans. See *Corone.*

Black Russian. See *Black Tartarian.*

Black Tartarian (*Black Circassian; Black Russian; Fraser's Black; Ronalds' Black; Sheppard's Seedling*). —Large, obtuse heart-shaped; surface irregular and uneven. Skin deep black. Stalk an inch and a half long. Flesh purplish, juicy, half tender, and rich. Stone small, roundish oval. Succeeds well against a wall, when it is ready by the end of June.

Bleeding Heart. See *Gascoigne's Heart.*

Bloodgood's Amber. See *American Amber.*

Bloodgood's Honey. See *American Amber.*

Bouquet Amarelle. See *Cluster.*

Bowyer's Early Heart.—Rather below medium size, obtuse heart-shaped. Skin amber coloured, mottled with red. Flesh white, very tender, juicy, and sweet. A good bearer, and an excellent early cherry. End of June.

Brant.—Large, roundish-heart-shaped, and uneven. Skin deep dark red. Stalk an inch and a half long, set in an angular cavity. Flesh dark purplish-red, half tender, juicy, sweet, and richly flavoured. Beginning of July.

Brune de Bruxelles. See *Ratafia.*

Bullock's Heart. See *Ox-Heart.*

BUTTNER'S BLACK HEART.—Larger than the common Black Heart; heart-shaped, and flattened on one side. Skin glossy, deep blackish-purple. Stalk an inch and a half long. Flesh half tender, juicy, dark red and with a particularly pleasant flavour. A superior variety to the common Black Heart. Ripe in the middle of July.

BUTTNER'S OCTOBER MORELLO.—Large, round, and somewhat oblate, and indented at the apex. Skin thin, and of a reddish-brown colour. Stalk slender, two inches long. Flesh light red, reticulated with whitish veins, juicy, and with a pleasant sub-acid flavour. This is an excellent culinary cherry, and ripens in October.

BUTTNER'S YELLOW.—About medium size, roundish-ovate. Skin entirely yellow, becoming a pale-amber colour when highly ripened. Stalk an inch and a half long. Flesh firm, yellow, sweet, and very nicely flavoured. The best yellow cherry there is. The birds do not touch it. Middle of July, and hangs till the end of August.

CARNATION (*Nouvelle d'Angleterre; De Villenne; Rouge pâle; Wax Cherry*).—Large, round, and flattened, inclining to oblate. Skin thin, light red at first, but changing to a deeper colour as it hangs; pale yellow, or amber, where shaded. Stalk an inch and a half long, and stout. Flesh pale yellow, rather firmer than in Dukes generally, juicy, and with a fine, brisk, sub-acid flavour, becoming richer the longer it hangs. Ripe in the end of July and beginning of August.

Cerise à Bouquet. See *Cluster*.

Cherry Duke *of Duhamel*. See *Jeffreys' Duke*.

CLEVELAND BIGARREAU (*Cleveland*).—Large, obtuse heart-shaped, sometimes with a swelling on one side near the stalk. Skin pale yellow, with bright red next the sun, and mottled with crimson. Stalk two inches long. Flesh yellowish white, half tender, juicy, sweet, and richly flavoured. Ripe the third or last week in June and early in July.

CLUSTER (*Cerise à Bouquet; Bouquet Amarelle; Flanders Cherry*).—Small, produced in a cluster of two, three, four, or five together at the end of one common stalk. Skin thin, at first of pale red, but changing to darker red the longer it hangs. Flesh white, tender, and juicy, very acid at first, but becoming milder as it hangs on the tree. Ripe from the middle to the end of July.

Coe's Late Carnation. — Medium sized, roundish. Skin reddish yellow clouded and mottled with bright red. Stalk two inches long. Flesh tender, juicy, with a brisk sub-acid flavour, becoming mellowed the longer it hangs. Ripe from the middle to the end of August, and continues to hang till September.

Cœur de Pigeon. See *Monstrous Heart.*

Common Red. See *Kentish.*

Corone (*Belcher's Black; Black Bud of Buckinghamshire; Black Corone; Black Orleans; Herefordshire Black*).—Small, roundish-heart-shaped. Skin deep blackish purple. Stalk two inches long, inserted in a deep, narrow cavity. Flesh dark purple, very firm, juicy, and sweet. Ripe in the end of July and beginning of August.

Dauphine. See *Belle de Choisy.*

Delicate.—Large, roundish, and flattened. Skin thin and translucent, fine rich amber coloured, quite covered with mottling of crimson. Stalk two inches long. Flesh pale yellow, translucent, tender, juicy, sweet, and with a rich, delicious flavour. A very excellent cherry. Ripe in the middle of July.

Doctor. See *American Doctor.*

Donna Maria. See *Royal Duke.*

Double Glass. See *Great Cornelian.*

Double Natte.—Rather large, roundish, a little compressed, and inclining to ovate. Skin dark brown, or brownish black. Stalk slender, sometimes nearly three inches long, and bearing leaves. Flesh very red, tender, and very juicy, with a brisk, sprightly acidity. Ripe in the beginning and middle of July.

Doucette. See *Belle de Choisy.*

Downer's Late.—Fruit produced in large bunches, medium sized, obtuse heart-shaped. Skin of a delicate clear red on the exposed side, but paler and mottled with pale yellow where shaded. Stalk an inch and a half long. Flesh pale, tender, juicy, sweet, and richly flavoured. Ripe the middle and end of August.

Downton.—Large, roundish-heart-shaped, much compressed, nearly round. Skin pale yellow, stained with red dots, semi-transparent, marbled with dark red next the sun. Stalk an inch and a half long, slender. Flesh yellowish, without any stain of red, tender, deliciously and richly flavoured, adhering slightly to the stone, Ripens in the middle and end of July.

Dredge's Early White. See *White Heart.*

Duchesse de Palluau.—Very large, oblate, and pitted at the apex. Skin thin, of a brilliant red colour, becoming dark red as it ripens. Stalk an inch and a half long. Flesh very tender and juicy, with a brisk and agreeable acidulous flavour; juice coloured. A very fine cherry. Ripe in the end of July.

Duke. See *May Duke.*

Dutch Morello. See *Morello.*

Early Amber (*Early Amber Heart; Rivers Early Amber Heart*).—Above medium size, heart-shaped. Skin pale amber, with a flush of red next the sun. Stalk two inches long. Flesh pale yellow, juicy, sweet, and richly flavoured. Beginning of July.

Early Black. See *Black Heart.*

Early Duke. See *May Duke.*

Early May (*Small May; Indulle; Nain Précoce*).—Small, round, slightly flattened. Skin lively light red. Stalk an inch long, slender, deeply set. Flesh soft, juicy, and acid. Middle of June.

Early May Duke. See *May Duke.*

Early Prolific.—Above medium size, obtuse-heart-shaped. Skin pale amber, mottled with crimson. Stalk two inches long. Flesh tolerably firm, juicy, rich, sweet, and delicious. End of June.

Early Purple Gean (*Early Purple Griotte; German May Duke*).—Large, obtuse heart-shaped, slightly flattened on one side. Skin shining, dark purple, almost black. Stalk slender, from two to two and a half inches long. Flesh dark purple, tender, and very juicy, with a very sweet and rich flavour. Ripe in the middle of June.

Early Purple Griotte. See *Early Purple Gean.*

Early Richmond. See *Kentish.*

Elkhorn. See *Tradescant's Heart.*

Elton.—Large, and heart-shaped. Skin thin, pale yellow in the shade, but mottled and streaked with bright red next the sun. Stalk two inches long, slender. Flesh half tender, juicy, very rich and luscious. Early in July.

Flanders. See *Cluster.*

FLEMISH.—Pomologists have fallen into great mistakes with regard to this cherry, particularly those who make it synonymous with Gros Gobet; others think it the same as the Kentish. The latter is nearer the truth; but the Kentish and Flemish are decidedly different. The fruit of the two could not be distinguished the one from the other; but the trees of the Flemish are less drooping than those of the Kentish, and the fruit is smaller, and about eight or ten days later. Anyone who examines the two varieties as they are grown in the Kentish orchards will see at once that the varieties are different.

FLORENCE (*Knevett's Late Bigarreau*).—Large and obtuse heart-shaped. Skin pale amber, marbled with red, and mottled with bright red where exposed. Stalk two inches long, slender, deeply set. Flesh yellowish, firm, very juicy, sweet, and rich. Beginning and middle of August.

Four-to-the-Pound. See *Tobacco-Leaved.*

Fraser's Black Tartarian. See *Black Tartarian.*

Fraser's White Tartarian. See *White Tartarian.*

Fraser's White Transparent. See *White Tartarian.*

GASCOIGNE'S HEART (*Bleeding Heart; Herefordshire Heart; Red Heart*).—Above medium size, heart-shaped, broad at the stalk, and terminating at the apex in an acute, swollen point. Skin entirely covered with bright red. Stalk two inches long, slender. Flesh yellowish white, half-tender, juicy, and sweet. Beginning and middle of July.

German May Duke. See *Early Purple Gean.*

Gobet à Courte Queue. See *Gros Gobet.*

GOVERNOR WOOD.—Large, obtuse heart-shaped. Skin pale yellow, washed and mottled with bright red. Stalk an inch and a half long. Flesh half-tender, juicy, sweet, and very richly flavoured. Beginning of July.

Graffion. See *Bigarreau.*

GREAT CORNELIAN (*Double Glass*).—Very large, oblate, marked on one side with a very deep suture, which quite divides the fruit. Skin thin and translucent, at first of a light red, but becoming darker as it ripens. Stalk an inch and a half long. Flesh yellowish, tender, very juicy, with a fine sub-acid, vinous, and rich flavour. Beginning of July.

GRIOTTE DE CHAUX.—Large, roundish-oblate. Skin dark red, and shining. Stalk two inches long, and slender. Flesh dark, tender, melting, and very juicy, with a brisk sub-acid flavour. End of July.

GRIOTTE DE KLEPAROW (*Belle Polonaise*).—Medium sized, roundish-oblate. Skin dark red. Stalk two inches long. Flesh dark, tender, and juicy, with a rich, sweet, and sub-acid flavour. End of July.

GRIOTTE DE PORTUGAL—This is by some considered synonymous with the Archduke. It certainly bears a considerable resemblance to it in the size, form, and colour, of the fruit; but I have not yet had an opportunity of comparing trees of equal age, and growing under the same circumstances. I am, however, inclined to believe that they will prove to be, if not really identical, at least very similar.

Gros Cœuret. See *Monstrous Heart*.

GROS GOBET (*Montmorency; Montmorency à Courte Queue*).—Medium sized, oblate, marked on one side with a very deep suture, which forms quite a cleft at the stalk. Skin smooth and shining, of a fine clear red, but becoming darker as it ripens. Stalk very short and thick, half an inch to an inch long. Flesh white, tender, very juicy, and briskly acid; but when it hangs long it is agreeably flavoured. Middle and end of July. This has been, very incorrectly, made synonymous with the Flemish, and even with the Kentish.

Grosse de Wagnelée. See *Reine Hortense*.

Guigne Noire Tardive. See *Tradescant's Heart*.

HARRISON'S HEART (*White Bigarreau*).—Very large, distinctly heart-shaped, and uneven in its outline. Skin at first of a yellowish white, but becoming all over mottled with red. Stalk an inch and a half long. Flesh firm, but less so than the Bigarreau, rich, and deliciously flavoured. Middle and end of July.

Herefordshire Black. See *Corone*.

Herefordshire Heart. See *Gascoigne's Heart*.

Hogg's Black Gean.—Medium sized, obtuse heart-shaped. Skin black and shining. Stalk an inch and a half long. Flesh dark, very tender, richly flavoured, and very sweet. Beginning of July.

Hogg's Red Gean.—Medium sized, roundish, inclining to heart-shaped. Skin red, freckled with amber yellow. Stalk an inch and a half long. Flesh yellowish, very tender and juicy, sweet, and richly flavoured. Beginning of July.

Hybrid de Laeken. See *Reine Hortense.*

Indulle. See *Early May.*

Jeffreys' Duke (*Cherry Duke* of Duhamel; *Jeffreys Royal; Royale*).—Medium sized, round, and a little flattened, produced upon stalks of about an inch long, which are united in clusters on one common peduncle half an inch long. Skin deep red, changing to black as it attains maturity. Flesh red, firm, very juicy, rich, and highly flavoured. The juice is quite sweet, and not acid, like the May Duke. Ripe the beginning and middle of July.

Joc-o-sot.—Large and handsome, somewhat obtusely heart-shaped, compressed on the sides, and deeply indented at the apex. Skin shining, of a deep brownish-black colour. Stalk two inches long. Flesh dark brownish-red, tender, juicy, rich, and sweet. Middle of July.

Kennicott.—Large, roundish-heart-shaped, and compressed on the sides. Skin of a fine amber yellow, considerably mottled with deep glossy red. Flesh yellowish white, firm, juicy, rich, and sweet. Beginning and middle of August.

Kentish (*Common Red; Early Richmond; Pie Cherry; Sussex; Virginian May*).—Medium sized, round. Skin bright red. Stalk an inch and a quarter long, stout, deeply set, and adhering so firmly to the stone, that the latter may be drawn out. Flesh acid. For kitchen purposes. Middle and end of July.

Kirtland's Mammoth. See *Mammoth.*

Kirtland's Mary. See *Mary.*

Knevett's Late Bigarreau. See *Florence.*

Knight's Early Black.—Large, and obtuse heart-shaped, irregular, and uneven. Skin black. Stalk two inches long, deeply inserted. Flesh purple, tender, juicy, and richly flavoured. End of June and beginning of July.

Lacure. See *Black Heart.*

Lady Southampton's.—This is a medium sized, yellow, heart-shaped cherry, of the Bigarreau class, with firm, but not juicy, flesh. It is now very little cultivated, and is but a worthless variety. End of July and beginning of August.

Large Black Bigarreau. See *Tradescant's Heart.*

Late Bigarreau.—Large, obtuse heart-shaped, and uneven in its outline, broadly and deeply indented at the apex. Skin of a fine rich yellow, with a bright red cheek, which sometimes extends over the whole surface. Stalk an inch and a half long. Flesh yellowish, considerably firm, sweet, and agreeably flavoured. Middle of August. Tree very productive.

Late Duke (*Anglaise Tardive*).—Large, obtusely heart-shaped, and somewhat compressed. Skin shining, of a fine bright red, which becomes darker as it ripens. Stalk one inch and a half to two inches long. Flesh pale yellow, tender, juicy, and richly flavoured. Beginning and middle of August.

Late Morello. See *Morello.*

Lemercier. See *Reine Hortense.* There is a Lemercier grown by Mr. Rivers which is later than Reine Hortense, and, before it is quite ripe, considerably more acid than that variety. The tree has also a more rigid and upright growth, like the Dukes; but it is evidently a seminal variety of Reine Hortense, and, being a better bearer, is, perhaps, the preferable kind to grow of that admirable cherry.

Lion's Heart. See *Ox Heart.*

Logan.—Above medium size, obtuse heart-shaped. Skin deep blackish-purple. Stalk an inch and a half long. Flesh brownish-red, almost firm, juicy, sweet, and richly flavoured. Middle and end of July. The tree blooms late.

Louis XVIII. See *Reine Hortense.*

Luke Ward's (*Lukewards*).—Medium sized, obtuse heart-shaped. Skin dark brownish-red, becoming almost black as it ripens. Stalk about two inches long. Flesh half-tender, dark purple, juicy, sweet, and richly flavoured. End of July and beginning of August. Superior to the Black Heart and the Corone.

MAMMOTH (*Kirtland's Mammoth*).—Very large, often an inch and an eighth in diameter; obtuse heart-shaped. Skin clear yellow, flushed and marbled with red. Stalk an inch and a quarter long. Flesh half-tender, juicy, sweet, and very richly flavoured. Middle and end of July. This is a magnificent cherry.

MANNING'S MOTTLED.—Above medium size, obtusely heart-shaped, and flattened on one side. Skin amber coloured, finely mottled, and flushed with red, somewhat translucent and shining. Stalk slender, two inches long. Flesh yellow, tender, juicy, sweet, and richly flavoured. Middle of July.

Marcelin. See *Monstrous Heart*

MARY (*Kirtland's Mary*). — Large, roundish-heart-shaped, and handsome. Skin very much mottled with deep, rich red on a yellow ground, and, when much exposed to the sun, almost entirely of a rich glossy red. Stalk from one inch and a half to two inches long. Flesh pale yellow, firm, rich, and juicy, with a sweet and high flavour. Middle and end of July. This is a very beautiful and very fine cherry.

MAY DUKE (*Duke; Early Duke; Early May Duke; Angleterre Hâtive; Royale Hâtive*).—Large, roundish, inclining to oblate. Skin at first of a red-cornelian colour, but gradually becoming dark red, and ultimately almost black, as it ripens. Stalk about an inch and a half long. Flesh red, tender, juicy, and richly flavoured, with a fine, subdued acidulous smack. Beginning of July.

De Meruer. See *Reine Hortense.*

Merveille de Hollande. See *Reine Hortense.*

Merveille de Septembre. See *Tardive de Mans*

Milan. See *Morello.*

MONSTROUS HEART (*Bigarreau Gros Cœuret; Bigarreau Jaboulais; Bigarreau Gaboulais; Bigarreau de Lyons; Bigarreau Gros Monstrueux; Bigarreau Monstrueux de Mezel; Marcelin; Gros Cœuret; Ward's Bigarreau*).—Very large, obtuse heart-shaped. Skin at first yellowish, tinged and streaked with red, but changing to a deep, shining red, and approaching to black the longer it hangs. Stalk one inch and a half to two inches long, stout. Flesh purplish, firm, and juicy, with a rich and excellent flavour. Ripe the middle and end of July.

Monstrueuse de Bavay. See *Reine Hortense.*

Monstrucuse de Jodoigne. See *Reine Hortense.*

Montmorency. See *Gros Gobet.*

Montmorency à Courte Queue. See *Gros Gobet.*

MORELLO (*Black Morello; Dutch Morello; Late Morello; Ronalds' Large Morello; Milan*).—Large, roundish, inclining to heart-shape. Skin dark red, becoming almost black the longer it hangs. Stalk an inch and a half to two inches long. Flesh purplish red, tender, juicy, and pleasantly sub-acid. Used for culinary purposes. July and August.

Morestein. See *Reine Hortense.*

Nain Précoce. See *Early May.*

Nouvelle d'Angleterre. See *Carnation.*

OHIO BEAUTY.—Large, obtuse heart-shaped. Skin pale yellow, overspread with red. Flesh pale, tender, brisk, and juicy. Beginning of July.

OSCEOLA.—Above medium size, heart-shaped, and with a deep suture on one side. Skin dark purplish-red, almost black. Stalk about two inches long. Flesh liver coloured, tender, very juicy, rich, and sweet. Middle and end of July.

OSTHEIM.—Large, roundish-oblate, compressed on one side. Skin red, changing to very dark red as it ripens. Stalk an inch and a half to two inches long. Flesh dark red, tender, and juicy, with a pleasant, sweet, and sub-acid flavour. An excellent preserving cherry, not so acid as the Morello. End of July.

Ounce Cherry. See *Tobacco-Leaved.*

OX HEART (*Bullock's Heart; Lion's Heart*).—Large, obtuse heart-shaped, flattened on one side. Skin shining, dark purplish-red. Stalk two inches long. Flesh somewhat firm, dark red, with a brisk and pleasant flavour, which is considerably richer when the fruit is highly ripened. End of July.

PARAMDAM (*Baramdam*).—Small and round, not quite

half an inch in diameter. Skin pale red. Stalk an inch long. Flesh pale, tender, with an agreeable and lively acidity. End of July. The tree is of very diminutive growth; one in my possession, not less than 100 years old, being little more than seven feet high, and the stem not so thick as a man's arm.

De Palembre. See *Belle de Choisy.*

Pie Cherry. See *Kentish.*

PONTIAC.—Large, obtuse heart-shaped, compressed on the sides. Skin dark purplish-red, nearly black. Stalk an inch and a half to two inches long. Flesh purplish red, half-tender, juicy, sweet, and agreeable. The latter end of July.

POWHATTAN. — Medium sized, roundish-heart-shaped, compressed on the sides, uneven in its outline. Skin brownish red and glossy. Stalk two inches long. Flesh rich purplish-red, half-tender, juicy, sweet, but not highly flavoured. End of July.

Quatre à la Livre. See *Tobacco-Leaved.*

RATAFIA (*Brune de Bruxelles*).—Medium sized, round, and a little flattened at both ends. Skin dark brown, nearly black, and very shining. Stalk an inch and a half to two inches long. Flesh dark red, tender, and juicy, with a brisk acid flavour, which becomes subdued the longer it hangs on the tree. August.

Red Heart. See *Gascoigne's Heart.*

REINE HORTENSE (*D'Aremberg; Belle Audigeoise; Belle de Bavay; Belle de Laeken; Belle de Prapeau; Belle de Petit Brie; Belle Suprême; Grosse de Wagnelée; Hybrid de Laeken; Louis XVIII.; Lemercier; De Meruer; Merveille de Hollande; Monstrueuse de Bavay; Monstrueuse de Jodoigne; Morestein; Rouvroy; Seize à la Livre*).—Very large, one inch and one-twelfth long and an inch wide, oblong, and compressed on the sides. Skin very thin and translucent, at first pale red, but assuming a bright cornelian red, and changing to dark brilliant red the longer it hangs. Stalk very slender, about two inches long. Flesh yellow, netted, very tender, and very juicy, with a sweet and agreeably acidulous juice. Middle of July.

RED JACKET. — Large, heart-shaped. Skin amber, covered with pale red, but when fully exposed entirely covered with bright red. Stalk two inches long, slender. Flesh half-tender, juicy, and of good, but not high, flavour. Beginning and middle of August. Valuable for its lateness.

Rivers' Early Amber Heart. See *Early Amber*.

ROCKPORT BIGARREAU.—Large, obtuse heart-shaped, uneven in its outline, and with a swelling on one side. Skin pale amber, covered with brilliant deep red, mottled and dotted with carmine. Stalk an inch to an inch and a half long. Flesh yellowish white, firm, juicy, sweet, and richly flavoured. Beginning and middle of July.

Ronalds' Black. See *Black Tartarian*.

Ronalds' Large Morello. See *Morello*.

Rouge Pâle. See *Carnation*.

Rouvroy. See *Reine Hortense*.

ROYAL DUKE (*Donna Maria*).—Large, oblate, and handsomely shaped. Skin deep, shining red, but never becoming black, like the May Duke. Stalk an inch and a half long. Flesh reddish, tender, juicy, and richly flavoured. Middle of July.

Royale. See *Jeffreys' Duke*.

Royale Hâtive. See *May Duke*.

St. Margaret's. See *Tradescant's Heart*.

Seize à la Livre. See *Reine Hortense*.

SHANNON MORELLO.—Above medium size, round, and flattened at the stalk. Skin dark purplish red. Stalk long and slender. Flesh tender, reddish purple, juicy, and acid. August.

Sheppard's Bedford Prolific. See *Black Tartarian*.

Small May. See *Early May*.

DE SOISSONS.—Medium sized, roundish, inclining to heart-shaped, and somewhat flattened at the apex. Skin dark red. Stalk not more than an inch long. Flesh red, tender, and juicy, with a brisk and pleasant sub-acid flavour. A good cherry for culinary purposes. Ripe in the middle and end of July.

Spanish Heart. See *Black Heart*.

SPARHAWK'S HONEY (*Sparrowhawk's Honey*).—Medium sized, obtuse heart-shaped, and very regular in shape. Skin thin, of a beautiful, glossy, pale amber, becoming a lively red when fully ripe, and somewhat translucent. Stalk of moderate length, rather slender. Flesh pale, juicy, and sweet. Middle of July.

Spotted Bigarreau. See *Bigarreau de Hollande.*

Sussex. See *Kentish.*

TARDIVE DE MANS (*Merveille de Septembre*).—Small, ovate, flattened at the stalk. Skin smooth and shining, clear red in the shade, and mottled with purplish red where exposed. Flesh firm, sweet, juicy, and nicely flavoured. This, like Belle Agathe, hangs very late, but it is not so large nor so good as that variety.

TECUMSEH.—Above medium size, obtuse heart-shaped, flattened on one side. Skin reddish purple, or dark brownish-red, mottled with red. Flesh reddish purple, half-tender, very juicy and sweet, but not highly flavoured. Middle and end of August. Valuable as a late variety.

TOBACCO-LEAVED (*Four-to-the-Pound; Ounce Cherry; Quatre à la Livre*).—Rather below medium size, heart-shaped, and somewhat flattened on one side, and terminating at the apex in a curved fleshy point. Skin thin, pale amber, mottled and spotted with red. Stalk slender, two inches long. Flesh pale-amber coloured, firm, juicy, and with a sweet, rich flavour. Beginning of August. Leaves nearly a foot long.

TRADESCANT'S HEART (*Elkhorn; St. Margaret's; Large Black Bigarreau; Bigarreau Gros Noir; Guigne Noire Tardive*).—Of the largest size, obtuse heart-shaped, indented and uneven on its surface, and considerably flattened next the stalk; on one side marked with the suture. Skin at first dark red, but changing when fully ripe to dark blackish-purple. Stalk slender, an inch and a half to an inch and three quarters long. Flesh dark purple, adhering firmly to the stone, firm, sweet, and briskly sub-acid. End of July and beginning of August.

TRANSPARENT GEAN.—Small, regularly heart-shaped. Skin thin, transparent, and shining, pale yellow, and finely mottled with clear red. Stalk two inches long, slender. Flesh pale, tender, and juicy, with a sweet and somewhat piquant flavour. Middle and end of July.

Trempée Précoce. See *Baumann's May.*

De Villenne. See *Carnation.*

Virginian May. See *Kentish.*

Ward's Bigarreau. See *Monstrous Heart.*

WATERLOO.—Large, obtuse heart-shaped, depressed at the apex, and flattened on one side. Skin very dark reddish-purple, almost black, and covered with minute pale dots. Stalk an inch and a half long, very slender. Flesh light reddish-purple, but dark purple next the stone; tender and juicy, with a sweet and rather rich flavour. End of June and beginning of July.

Wax Cherry. See *Carnation.*

WERDER'S EARLY BLACK.—Very large, obtuse heart-shaped, with a deep suture on one side. Skin tough, shining, deep black-purple. Stalk short and stout, about an inch and a half long. Flesh purplish red, tender, very juicy, and with a very sweet and rich flavour. Middle and end of June.

West's White Heart. See *Bigarreau.*

White Bigarreau. See *Harrison's Heart.*

WHITE HEART (*Amber Heart; Dredge's Early White; White Transparent*).—Above medium size, oblong heart-shaped. Skin whitish yellow, tinged with dull red next the sun. Stalk two inches long, slender, set in a wide cavity. Flesh half-tender, sweet, and pleasant. Stone large. End of July.

WHITE TARTARIAN (*Fraser's White Tartarian*).—Medium sized, obtuse heart-shaped. Skin pale yellow. Stalk two inches long, slender. Flesh whitish yellow, half-tender, and sweet. Early in July.

White Transparent. See *White Heart.*

---

## LIST OF SELECT CHERRIES,

### ARRANGED ACCORDING TO THEIR ORDER OF RIPENING.

### I. FOR GARDENS.

These all succeed well in the open ground, or as espaliers; and those for dessert use are all worthy of being grown against a wall, when they are much improved both in quality and earliness.

### *For Dessert Use.*

JUNE.

Belle d'Orléans
Early Purple Gean
Baumann's May
Early Prolific
Werder's Early Black
Bowyer's Early Heart

JULY.

Knight's Early Black
Black Tartarian
Waterloo
Governor Wood
Belle de Choisy
May Duke
Jeffreys' Duke
Cleveland Bigarreau
Rockport Bigarreau
Black Eagle
Elton
Osceola
Royal Duke
Delicate
Duchesse de Palluau
Monstrous Heart
Joc-o-sot
Mammoth
Mary
Bigarreau

AUGUST.

Late Duke
Florence
Kennicott
Red Jacket
Tecumseh

SEPTEMBER.

Coe's Late Carnation
Büttner's Yellow
Bigarreau de Hildesheim
Belle Agathe

### *For Kitchen Use.*

Kentish
Griotte de Chaux
Belle Magnifique
Morello

---

## II. FOR ORCHARDS.

These being vigorous-growing and hardy varieties, and all, in various degrees, abundant bearers, are well adapted for orchard planting.

Early Prolific
Knight's Early Black
Black Tartarian
Adams' Crown
May Duke
Elton
Black Hawk
Büttner's Black Heart
Hogg's Black Gean
Hogg's Red Gean
Kentish
Mammoth
Mary
Bigarreau
Amber Gean
Late Duke
Kennicott
Red Jacket
Tecumseh
Belle Agathe

## CHESTNUTS.

We can hardly call the chestnut a British fruit. It is true, that in some situations in the southern counties it ripens fruit, but that is generally so very inferior to what is imported from Spain and the South of France, that no one would think of planting the chestnut for its fruit alone. It is as a timber tree that it is so highly valued in this country.

The following are the varieties that succeed best; but it is only in hot summers that they attain much excellence:—

Devonshire Prolific (*New Prolific*).—This is by far the most abundant bearer, and ripens more thoroughly a general crop than any other.

Downton (*Knight's Prolific*).—This is distinguished by the very short spines on the husks, and is not so prolific as the preceding.

## CRABS.

These are grown mainly for ornament. Their fruit, being generally very highly or delicately coloured, contribute to the decoration of shrubberies in the autumn; while their flowers make them gay with blossoms in the spring. But there are some of the varieties which, besides being ornamental, are also very useful for preserving. Of these, the following are the most esteemed:—

Cherry Crab (*Cherry Apple; Scarlet Siberian*).—Very small, the size and colour of a cherry, roundish oblong, flat at the ends, of a bright shining scarlet colour, with the appearance as if it had been varnished. Stalk very long and slender. Eye small. Flesh crisp, with a fine agreeable acidity. Used for preserving. September and October.

Royal Charlotte.—Medium sized, ovate. Skin of a delicate waxen yellow, tinged with red all over, but covered with a dark red cheek next the sun. Eye with long, pointed segments, and moderately sunk. Stalk slender, an inch long. Flesh white, very tender, with a fine, agreeable acidity. September and October.

Siberian (*Yellow Siberian*).—Small, conical. Skin waxen yellow in the shade, and streaked with red next the sun. Eye large and protruding, closed. Flesh briskly acid. September and October.

Transparent.—Below medium size, oblate. Skin yellowish white, and waxen-like. Eye with very long and spreading segments, sunk. Stalk long and slender. Flesh translucent, opaline, with a brisk and agreeable acidity. October.

## CRANBERRIES.

THOUGH these cannot be grown so generally as the other kinds of fruits, there are some who, having devoted their attention to the subject, have succeeded in forming artificial swamps where cranberries have been cultivated with great success. Wherever there is a command, and a plentiful supply of running water, with abundance of peat soil, no difficulty need be experienced in growing cranberries. The two species most worth cultivating are the English and the American.

ENGLISH (*Oxycoccus palustris*).—This grows abundantly in bogs, or swamps, in many parts of England. The fruit is the size of a pea, and the skin pale red; they have a somewhat acrid flavour and a strong acidity.

AMERICAN (*Oxycoccus macrocarpus*).—Of this there are three varieties :—

1. *Cherry Cranberry*, is large, round, and of a dark red colour, resembling a small cherry.

2. *Bugle Cranberry*, so called from the shape being like a bugle bead, long, and approaching an oval. Skin pale, and not so deep a crimson as the other varieties.

3. *Bell Cranberry*, is bell-shaped, or turbinate, and of a dark coral red. This is a very large variety, and is a great favourite with American growers.

## CURRANTS.

Black Grape. See *Ogden's Black.*

BLACK NAPLES (*New Black*).—Bunches short, but produced in great abundance. Berries larger than any other variety, frequently measuring about three quarters of an inch in diameter. Milder and sweeter than any other black currant, and the best of all the black varieties.

Cerise. See *Cherry.*

CHAMPAGNE (*Pheasant's Eye; Couleur de Chair*).—Bunches of medium length. Berries medium sized, pale pink, or flesh coloured, with darker red veins; more acid than Red Dutch.

CHERRY (*Cerise*).—Bunches short. Berries very large, of a deep red colour; more acid than Red Dutch. This is the largest red currant, and comes in early.

COMMON BLACK.—This is very much inferior to Black Naples and Ogden's Black, and not worth cultivation, the bunches and berries being inferior in size to both of those varieties.

Couleur de Chair. See *Champagne.*

Goliath. See *Raby Castle.*

Houghton Castle. See *Raby Castle.*

Jeeves' White. See *White Dutch.*

KNIGHT'S EARLY RED.—The chief merit this variety is supposed to possess, is its greater earliness than the Red Dutch; but the slight advantage it has in this, is lost by its inferiority in other respects.

KNIGHT'S LARGE RED. — Bunches large and long. Berries large, bright red. Does not differ materially from Red Dutch.

KNIGHT'S SWEET RED. — Bunches of medium size. Berries large, paler in colour than Red Dutch, and less acid; but not so sweet as White Dutch.

LA FERTILE.—This variety I have not seen; but, according to Mr. Rivers, it is a large red currant, and "a most prodigious bearer."

LA HÂTIVE.—This is a new variety, and, like the preceding, of foreign origin; but I have had no opportunity of examining it. Mr. Rivers states, in his catalogue, that it is "a very early red currant, and excellent."

LONG-BUNCHED RED (*Wilmot's Long-bunched Red*).—Bunches very long, sometimes measuring six inches and a half. Berries large, and of a deep red colour. A decided improvement on Red Dutch, and differs also in being somewhat later. It is not unlike Raby Castle.

May's Victoria. See *Raby Castle.*

Morgan's White. See *White Dutch.*

New Black. See *Black Naples.*

New White Dutch. See *White Dutch.*

OGDEN'S BLACK (*Black Grape*).—This is not so large as Black Naples, but considerably better in every respect than the Common Black. The bush is hardier than that of Black Naples.

Pheasant's Eye. See *Champagne.*

RABY CASTLE (*Houghton Castle; May's Victoria; Victoria; Goliath*).—Bunches longer than those of Red Dutch; berries larger, and of a brighter red, but rather more acid. It is an abundant bearer, and the fruit ripens later, and hangs longer, than any other currant.

RED DUTCH (*Large Red Dutch; New Red Dutch; Red Grape*).—Bunches from two to three inches long. Berries large, deep red, with a subdued acidity. Superior in every respect to the old Common Red, which is unworthy of cultivation.

Red Grape. See *Red Dutch.*

White Crystal. See *White Dutch.*

WHITE DUTCH (*New White Dutch; Jeeves' White; Morgan's White; White Crystal; White Leghorn; White Grape*).—The bunches and berries are of the same size as the Red Dutch; but the berries are of a yellowish white, and the skin somewhat transparent. The fruit is very much sweeter, and more agreeable to eat, than the Red variety. It is, therefore, preferred in the dessert, and for wine-making.

White Grape. See *White Dutch.*

White Leghorn. See *White Dutch.*

Wilmot's Long-bunched Red. See *Long-bunched Red.*

## LIST OF SELECT CURRANTS.

BLACK.

Black Naples
Ogden's Black

RED.

Cherry
Knight's Large Red
Long-bunched Red
Raby Castle
Red Dutch

WHITE.

White Dutch

# FIGS.

## SYNOPSIS OF FIGS.

### I. FRUIT ROUND, ROUNDISH, OR TURBINATE.

§ *Skin dark. Flesh red.*

Black Bourjassotte
Black Genoa
Black Ischia
Brown Ischia
Early Violet
Malta
Pregussata

§§ *Skin pale.*

* *Flesh red.*

Large White Genoa
Savantine
White Bourjassotte
White Ischia
Yellow Ischia

** *Flesh white.*

Angélique
Early White
Marseilles
Raby Castle

### II. FRUIT LONG, PYRIFORM, OR OBOVATE.

§ *Skin dark. Flesh red.*

Black Provence
Bordeaux
Brown Turkey
Brunswick
Peau dure
Violette Grosse

---

ANGÉLIQUE (*Mélitte; Madeleine; Coucourelle Blanche*).—Below medium size, about two inches long and an inch and three quarters broad; obovate. Skin yellow, dotted with long greenish-white specks. Flesh white under the skin, but tinged with red towards the centre. When well ripened, the fruit is of good quality, and perfumed. It requires artificial heat to bring it to perfection, and forces well.

Ashridge Forcing. See *Brown Turkey.*

D'Athènes. See *Marseilles.*

Aubique Violette. See *Bordeaux.*

Aubiquon. See *Bordeaux.*

Aulique. See *Violette Grosse.*

Barnissotte. See *Black Bourjassotte.*

Bayswater. See *Brunswick.*

De Bellegarde. See *Black Bourjassotte.*

Black Bourjassotte (*Précoce Noire; Barnissotte; De Bellegarde*).—Large, roundish. Skin dark purple. Flesh red at the centre, and of good quality; but requires heat to bring it to perfection. September. Tree an abundant bearer.

Black Genoa (*Nigra; Negro d'Espagne; Noire de Languedoc*).—Large, oblong, broad towards the apex, and very slender towards the stalk. Skin dark purple, almost black, and covered with a thick blue bloom. Flesh yellowish under the skin, but red towards the interior, juicy, with a very sweet and rich flavour. End of August. Tree very hardy, and a good bearer.

Black Ischia (*Blue Ischia; Early Forcing; Ronde Noire; Nero*).—Medium sized, turbinate, flat at the top. Skin deep purple, almost black when ripe. Flesh deep red, sweet, and luscious. Tree hardy, and an excellent bearer; succeeds well in pots. August.

Black Marseilles. See *Black Provence.*

Black Naples. See *Brunswick.*

Black Provence (*Black Marseilles*).—Below medium size, oblong. Skin dark brown. Flesh red, tender, very juicy, and richly flavoured. Tree bears abundantly, and is well adapted for forcing.

Blanche. See *Marseilles.*

Blue. See *Brown Turkey.*

Blue Burgundy. See *Brown Turkey.*

Blue Ischia. See *Black Ischia.*

Bordeaux (*Violette; Violette Longue; Violette de Bordeaux; Aubiquon; Aubique Violette; Petite Aubique; Figue-Poire*).—Large, pear-shaped, rounded at the head, and tapering to a small point at the stalk. Skin deep violet, strewed with long green specks. Flesh red, sweet, and well flavoured. Only a second-rate variety, and the tree is so tender, that it is apt to be cut down, even to the ground, by severe frosts.

Bourjassotte Blanche. See *White Bourjassotte.*

Bourjassotte Noire. See *Black Bourjassotte.*

Brocket Hall. See *White Ischia.*

Brown Hamburgh. See *Brunswick.*

Brown Ischia (*Chestnut-coloured Ischia*).—Medium sized, roundish-turbinate. Skin light brown, or chestnut coloured. Eye very large. Flesh purple, sweet, and high-flavoured. Fruit apt to burst by too much wet. This is one of the best of figs, ripening in the beginning and middle of August. Tree an excellent bearer, pretty hardy, and bears as a standard in favourable situations. It forces well.

Brown Italian. See *Brown Turkey.*

Brown Naples. See *Brown Turkey.*

Brown Turkey (*Ashridge Forcing; Blue; Common Blue; Blue Burgundy; Brown Italian; Brown Naples; Long Naples; Early; Howick; Italian; Jerusalem; Large Blue; Lee's Perpetual; Murrey; Purple; Small Blue; Fleur Rouge; Walton*).—Large and pyriform. Skin brownish red, covered with blue bloom. Flesh red and very luscious. Tree very prolific, hardy, and one of the best for out-door culture, as a standard. August and September.

Brunswick (*Bayswater; Black Naples; Brown Hamburgh; Clémentine; Hanover; Madonna; Large White Turkey; Rose Blanche; Rose Beyronne; Peronne; Rose; Red*).—Very large and pyriform, oblique at the apex, which is very much depressed. Skin greenish yellow in the shade; violet brown on the other side. Flesh yellow under the skin, tinged with red towards the centre. Very rich and excellent. Middle of August. The tree is very hardy and an excellent bearer, and certainly the best for out-door cultivation against walls.

Chestnut-coloured Ischia. See *Brown Ischia.*

Clémentine. See *Brunswick.*

Common Purple. See *Brown Turkey.*

Cyprus. See *Yellow Ischia.*

Early. See *Brown Turkey.*

Early Purple. See *Black Ischia.*

EARLY VIOLET.—Small, roundish. Skin brownish red, covered with blue bloom. Flesh red, and well flavoured. August. Tree hardy, and an abundant bearer; well adapted for pots and for forcing, when, according to Mr. Rivers, it bears three crops in one season.

EARLY WHITE (*Small White; Small Early White*).—Fruit roundish-turbinate, somewhat flattened at the apex. Skin thin, pale yellowish white. Flesh white, sweet, but not highly flavoured. August.

Figue-Poire. See *Bordeaux.*

Fleur Rouge. See *Brown Turkey.*

Ford's Seedling. See *Marseilles.*

Hanover. See *Brunswick.*

Howick. See *Brown Turkey.*

Italian. See *Brown Turkey.*

Jerusalem. See *Brown Turkey.*

Large Blue. See *Brown Turkey.*

LARGE WHITE GENOA.—Large, roundish-turbinate. Skin thin, of a pale yellowish colour, when fully ripe. Flesh red throughout, and of excellent flavour. End of August. This is a variety of first-rate excellence, but the tree is a bad bearer.

Large White Turkey. See *Brunswick.*

Lee's Perpetual. See *Brown Turkey.*

Long Naples. See *Brown Turkey.*

Madeleine. See *Angelique.*

Madonna. See *Brunswick.*

MALTA (*Small Brown*).—Small, roundish-turbinate, compressed at the apex. Skin pale brown, when fully ripe. Flesh the same colour as the skin; very sweet, and well flavoured. End of August. If allowed to hang till it shrivels, it becomes quite a sweetmeat.

Marseillaise. See *Marseilles.*

MARSEILLES (*Ford's Seedling; Pocock's; White Marseilles; White Naples; White Standard; D Athènes; Blanche; Marseillaise*).—Medium sized, roundish-turbinate, slightly depressed, and ribbed. Skin yellowish white. Flesh white, very melting and juicy, with a rich, sugary flavour. Ripe in August. One of the best for forcing; and also succeeds well in the open air against a wall.

Murrey. See *Brown Turkey.*

Negro d'Espagne. See *Black Genoa.*

Nerii. See *White Ischia.* There is no fig bearing this name distinct from White Ischia; and the variety Mr. Knight introduced under that designation was the same. By the name "Nerii," is intended the "Nero," or Black Fig, of the Italians, and the variety Mr. Knight received under that name was evidently incorrect; the true Fico Nero being the Black Ischia, and not the White Ischia.

Nero. See *Black Ischia.*

Noire de Languedoc. See *Black Genoa.*

Œil de Perdrix.—Small. Skin yellowish, with a brownish tinge, having a small, bright red circle under the surface round the eye: hence the origin of the name. Flesh white, tinged with red, rich, and highly flavoured. Tree an abundant bearer.

Peau Dure (*Peldure; Verte Brune*).—Medium sized, oblong ovate. Skin thick and tough, dark violet. Flesh purplish red, and well flavoured; but, when over-ripe, it acquires a little acerbity.

Peldure. See *Peau Dure.*

Peronne. See *Brunswick.*

Petite Aubique. See *Bordeaux.*

Pocock's. See *Marseilles.*

Précoce Noire. See *Black Bourjassotte.*

Pregussata.—Small, round, compressed at the ends. Skin purplish brown in the shade; dark brown, covered with pale spots, next the sun. Flesh deep red, rich, and luscious. August to October. Well adapted for forcing.

Purple. See *Brown Turkey.*

Raby Castle.—A variety closely resembling Marseilles, but distinguished from it by having a longer stalk.

Red. See *Brunswick.*

Ronde Noire. See *Black Ischia.*

Rose. See *Brunswick.*

Rose Beyronne. See *Brunswick.*

Rose Blanche. See *Brunswick.*

SAVANTINE (*Cordellière*).—Fruit round, marked along its length with prominent nerves. Skin pale yellow. Flesh pale red.

Singleton. See *White Ischia.*

Small Blue. See *Brown Turkey.*

Small Brown. See *Malta.*

Small Early White. See *Early White.*

Small White. See *Early White.*

Verte Brune. See *Peau Dure.*

Violette. See *Bordeaux.*

Violette de Bordeaux. See *Bordeaux.*

VIOLETTE GROSSE (*Aubique*).—Large, oblong, and perhaps the longest-shaped of any of the figs; its length being three times its diameter. Skin deep violet. Flesh red.

Violette Longue. See *Bordeaux.*

Walton. See *Brown Turkey.*

WHITE BOURJASSOTTE (*Bourjassotte Blanche*).—This is extensively cultivated about Marseilles. The fruit is turbinate. Skin yellowish white. Flesh red. The tree attains a large size.

WHITE ISCHIA (*Green Ischia; Nerii; Singleton; Brocket Hall*).—Small and turbinate. Skin pale greenish-yellow, very thin, so much so, that when fully ripe, the flesh, which is purple, shines through and gives the fruit a brownish tinge. Rich, highly flavoured, and luscious. End of August. The tree is of small habit of growth, a great bearer, well adapted for pot-culture, and forces well.

White Naples. See *Marseilles.*

White Standard. See *Marseilles.*

YELLOW ISCHIA (*Cyprus*).—Large, turbinate. Skin yellow. Flesh dark red, tender and very juicy, with a rich and sugary flavour. September.

## LIST OF SELECT FIGS.

### I. FOR STANDARDS.

Black Ischia
Brown Ischia
Brown Turkey

### II. FOR WALLS.

Black Genoa
Black Ischia
Brown Ischia
Brown Turkey
Brunswick
Marseilles

### III. FOR FORCING, OR POT-CULTURE.

Angélique
Black Ischia
Brown Ischia
Brown Turkey
White Ischia
Early Violet
Marseilles
Pregussata

## GOOSEBERRIES.

### SYNOPSIS OF GOOSEBERRIES.

#### I. SKIN RED.

§ *Round or Roundish.*

A. *Skin smooth.*

Prince Regent (Boardman's)
Small Red Globe

B. *Skin downy.*

Miss Bold
Scotch Nutmeg

C. *Skin hairy.*

Hairy Red (Barton's)
Irish Plum
Ironmonger
Lancashire Lad (Hartshorn's)
Raspberry
Rifleman (Leigh's)
Rough Red
Scotch Nutmeg
Shakespere (Denny's)
Small Rough Red
Top Sawyer (Capper's)
Victory (Lomas')

§§ *Oblong, oval, or obovate.*

A. *Skin smooth.*

Emperor Napoléon (Rival's)
Old England (Rider's)
Red Turkey
Ringleader (Johnson's)
Roaring Lion (Farrow's)
Sportsman (Chadwick's)
Wilmot's Early Red

B. *Skin downy.*

Farmer's Glory (Berry's)
Magistrate (Diggle's)
Red Walnut

C. *Skin hairy.*

Atlas (Brundrett's)
Beauty of England (Hamlet's)
Crown Bob (Melling's)
Early Black
Early Rough Red
Keens' Seedling
Over-All (Bratherton's)
Pastime (Bratherton's)
Red Champagne
Red Mogul
Red Oval
Red Warrington
Rob Roy
Yaxley Hero (Speechley's)

#### II. SKIN YELLOW.

§ *Round or Roundish.*

A. *Skin smooth.*

Amber
Yellow Ball

B. *Skin downy.*

Golden Drop
Rumbullion

C. *Skin hairy.*

Rockwood (Prophet's)
Sulphur
Yellow Champagne
Yellow Warrington

### §§ *Oblong, oval, or obovate.*

A. *Skin smooth.*

Duckswing (Buerdsill's)
Lord Combermere (Forester's)
Smiling Beauty (Beaumont's)
Victory (Mather's)
Viper (Gorton's)

B. *Skin downy.*

Husbandman (Foster's)
Invincible (Heywood's)
Prince of Orange (Bell's)

C. *Skin hairy.*

Conquering Hero (Catlow's)
Early Sulphur
Golden Fleece (Part's)
Golden Gourd (Hill's)
Yellowsmith

## III. SKIN GREEN.

### § *Round or Roundish.*

A. *Skin smooth.*

Glory of Kingston
Green Gage (Horsefield's)

B. *Skin downy.*

Green Willow
Joke (Hodkinsons's)
Perfection (Gregory's)

C. *Skin hairy.*

Green Gascoigne
Green Rumbullion
Hebburn Prolific

### §§ *Oblong, oval, or obovate.*

A. *Skin smooth.*

Favourite (Bates')
Glory of Ratcliff (Allen's)
Green Gage (Pitmaston)
Green Walnut
Heart of Oak (Massey's)
Independent (Briggs')
Jolly Tar (Edwards')

B. *Skin downy.*

Jolly Angler (Collier's)
Laurel (Parkinson's)
Profit (Prophet's)

C. *Skin hairy.*

Glenton Green
Wistaston Hero (Bratherton's)

## IV. SKIN WHITE.

### § *Round or Roundish.*

A. *Skin smooth.*

Crystal
White Rasp

B. *Skin downy.*

Early White

C. *Skin hairy.*

Hedgehog
Royal White
Snowball (Adams')

### §§ *Oblong, oval, or obovate.*

A. *Skin smooth.*

Lady Delamere (Wild's)
Lionness (Fennyhaugh's)
Queen Caroline (Lovart's)
White Eagle (Cooke's)
White Fig

| B. *Skin downy.* | C. *Skin hairy.* |
|---|---|
| Cheshire Lass (Saunders') | Abraham Newland |
| Sheba Queen (Crompton's) | Bonny Lass (Capper's) |
| Wellington's Glory | Bright Venus (Taylor's) |
| White Lily | Governess (Bratherton's) |
| White Lion (Cleworth's) | Lady of the Manor (Hopley's) |
| Whitesmith (Woodward's) | Princess Royal |
| | White Champagne |

---

THOSE varieties marked L.P. are of very large size, and are known as "Lancashire Prize Gooseberries." For the whole of these descriptions I am indebted to the Horticultural Society's Catalogue, as I have had no opportunity of personally examining this portion of the fruits of Great Britain.

ABRAHAM NEWLAND (Jackson's), L.P.—Large and oblong. Skin white and hairy. Highly flavoured and excellent. Bush erect.

AMBER (*Yellow Amber; Smooth Amber*). — Medium sized, roundish. Skin smooth, greenish yellow. Of good flavour, but not first-rate. Bush a good bearer; spreading.

Aston. See *Red Warrington.*

Aston Seedling. See *Red Warrington.*

ATLAS (Brundrett's), L.P.—Large, oblong. Skin red, hairy. Of good flavour, but not first-rate. Bush erect.

BEAUTY OF ENGLAND (Hamlet's), L.P.—Large and oblong. Skin red, hairy. Of good flavour. Bush spreading.

Belmont's Green. See *Green Walnut.*

BONNY LASS (Capper's), L.P.—Large, oblong. Skin white and hairy. Of second-rate quality. Bush spreading.

BRIGHT VENUS (Taylor's), L.P.—Medium sized, obovate. Skin slightly hairy, white, and covered with a bloom when it hangs long. Sugary, rich, and excellent, and hangs till it shrivels. Bush rather erect, and a good bearer.

British Prince. See *Prince Regent* (Boardman's).

CHAMPAGNE, RED (*Dr. Davies' Upright; Countess of Errol*).—Rather small and oblong, tapering a little towards the stalk. Skin rather thick, light red, and hairy.

Early. One of the richest flavoured of all the gooseberries; vinous, and very sweet. Bush very erect, and an excellent bearer. This is frequently, and in Scotland particularly, called "the Ironmonger."

CHESHIRE LASS (Saunders'), L.P.—Large and oblong. Skin very thin, downy, and white. Flavour rich and sweet. Bush erect, and a good bearer. Excellent for tarts, on account of its early attaining a size for that purpose.

CROWN BOB (Melling's), L.P.—Very large and oblong. Skin thin, hairy, bright red, with a greenish tinge toward the stalk. Of good flavour, and a first-rate variety. Bush pendulous, and an abundant bearer.

CRYSTAL.—Small and roundish. Skin thick, smooth, or very slightly downy, and white. Of good flavour, and chiefly valuable for coming in late. Bush spreading, and rather pendulous; leaves not hairy above.

Dr. Davies' Upright. See *Red Champagne.*

Double Bearing. See *Red Walnut* (Eckersley's).

DUCK WING (Buerdsill's), L.P. — Large and obovate. Skin yellow, and smooth. A late variety, and only of second-rate quality. Bush erect.

EARLY BLACK.—Medium sized, oblong. Skin dark red, and hairy. A second-rate variety. Bush pendulous.

EARLY ROUGH RED.—Small, roundish-oblong. Skin red, and hairy. A well-flavoured variety, but not first-rate. Bush spreading.

EARLY SULPHUR (*Golden Ball; Golden Bull; Moss's Seedling*).—Medium sized, roundish-oblong. Skin yellow, and hairy. Of second-rate quality. Bush erect, very early, and a great bearer; leaves downy.

EARLY WHITE.—Medium sized, roundish-oblong. Skin thin, transparent, yellowish white, and slightly downy. Very sweet, good, and early. A first-rate variety. Bush spreading and erect; an excellent bearer.

EMPEROR NAPOLÉON (Rival's), L.P.—Large and obovate. Skin red, and smooth. A second-rate variety. Bush pendulous, and a good bearer.

FARMER'S GLORY (Berry's), L.P.—Very large and obovate. Skin thick, downy, and dark red, with a mixture of green. A first-rate variety, and of excellent flavour. Bush pendulous, and an abundant bearer.

Favourite (Bates'). — Medium sized, oblong. Skin smooth, and green. Flavour second-rate. Bush pendulous.

Glenton Green (*York Seedling*). — Medium sized, oblong, narrowest at the base. Skin rather thick, very hairy, green, and with whitish veins. Of a sweet and an excellent flavour. Bush pendulous, and an excellent bearer. Young shoots downy, and sprinkled near the base with small prickles. Leaves downy above.

Glory of Kingston.—Medium sized, roundish. Skin smooth, and green. Not highly flavoured. Bush spreading, and a bad bearer.

Glory of Ratcliff (Allen's).—Medium sized, oblong. Skin thick, quite smooth, and light green. Of excellent flavour, and sweet. Bush spreading and somewhat pendulous, and a good bearer.

Golden Ball. See *Early Sulphur*.

Golden Bull. See *Early Sulphur*.

Golden Drop (*Golden Lemon*). — Medium sized, roundish. Skin downy, and yellow. Of second-rate quality. Bush erect.

Golden Fleece (Part's), L.P.—Very large, oval. Skin yellow, and hairy. Of first-rate quality.

Golden Gourd (Hill's), L.P.—Very large and oblong. Skin greenish yellow, and hairy. Of second-rate quality. Bush pendulous.

Golden Lemon. See *Golden Drop*.

Governess (Bratherton's), L.P. — Large, roundish-oblong. Skin greenish white, and hairy. Of second-rate quality. Bush spreading.

Green Gage (Horsefield's), L.P.—Large and roundish. Skin green and smooth. Flavour only third-rate. Bush spreading.

Green Gascoigne (*Early Green; Early Green Hairy*). —Small and round. Skin thin, dark green, and hairy. Very early, and sweet. Bush very erect, and an excellent bearer.

Green Laurel. See *Laurel*.

Green Walnut (*Belmont Green; Smooth Green, Nonpareil*).—Fruit medium sized, obovate. Skin very thin, dark green, and smooth. An early variet , of excellent flavour. Bush with long-spreading shoots; leaves close to the branches; and a great bearer.

Green Willow. See *Laurel.*

Grundy's Lady Lilford. See *Whitesmith* (Woodward's).

Hairy Amber. See *Yellow Champagne.*

Hairy Black. See *Ironmonger.*

HAIRY RED (Barton's).—Small and roundish. Skin thick, red, and slightly hairy. Briskly and well flavoured. Bush erect, and an excellent bearer.

Hall's Seedling. See *Whitesmith* (Woodward's).

HEART OF OAK (Massey's), L.P.—Large and oblong, tapering to the stalk. Skin thin, green, with yellowish veins. Rich and excellent. Bush pendulous, and an abundant bearer.

HEBBURN PROLIFIC.—Medium sized, roundish. Skin rather thick, dull green, and hairy. Very rich and sweet. Bush erect, with broad, thick leaves, and an abundant bearer.

HEDGEHOG. — Medium sized, roundish. Skin thin, white, and hairy. A richly-flavoured variety. Bush erect; the shoots thickly set with small briskly spines. This name is also applied to Glenton Green, in Scotland.

HUSBANDMAN (Foster's), L.P. — Large and obovate. Skin yellow, and downy. Of second-rate quality. Bush erect.

INDEPENDENT (Brigg's), L.P. — Large and obovate. Skin green and smooth. Of second-rate quality. Bush erect, and a good bearer.

INVINCIBLE (Heywood's), L.P.—Large and roundish-oblong. Skin yellow, and downy. Of second-rate quality. Bush erect.

IRISH PLUM.—Medium sized, roundish. Skin dark red and hairy. A first-rate dessert sort. Bush erect.

IRONMONGER (*Hairy Black*).—Small and roundish. Skin red, and hairy. A first-rate variety, of excellent flavour, but inferior to Red Champagne, which is also known under this name chiefly in Scotland; and from which it is distinguished in having rounder and darker red fruit, and a spreading bush—that of the Red Champagne being erect; leaves downy.

Jolly Anglers (Collier's), L.P. (*Lay's Jolly Angler*).—Large and oblong. Skin green, and downy. Of first-rate quality, and a good late sort. Bush erect.

Jolly Tar (Edwards'), L.P.—Large and obovate. Skin green, and smooth. Of first-rate quality. Bush pendulous, and a good bearer.

Keens' Seedling (*Keens' Seedling Warrington*).—Medium sized, oblong. Skin brownish red, hairy. Of first-rate quality. Bush pendulous; a great bearer, and earlier than Red Warrington.

Lancashire Lad (Hartshorn's), L.P. — Large and roundish. Skin dark red, and hairy. Of second-rate quality. Bush erect, and a good bearer.

Lancashire Lass. See *Whitesmith* (Woodward's).

Laurel (Parkinson's), L.P. (*Green Laurel; Green Willow*).—Large and obovate. Skin pale green, and downy. A first-rate variety, somewhat resembling Woodward's Whitesmith. Bush erect, and a good bearer.

Lay's Jolly Angler. See *Jolly Anglers* (Collier's).

Lord Combermere (Forester's), L.P.—Large and obovate. Skin yellow, and smooth. Of second-rate quality. Bush spreading.

Magistrate (Diggles'), L.P.—Large and obovate. Skin red, and downy. A first-rate variety. Bush spreading.

Miss Bold (*Pigeon's Egg*).—Medium sized, roundish. Skin red, and downy. Of first-rate quality, and early; it somewhat resembles Red Walnut, but is better. Bush spreading.

Moss' Seedling. See *Red Warrington.*

Murrey. See *Red Walnut.*

Nonpareil. See *Green Walnut.*

Nutmeg. See *Raspberry.*

Old England (Rider's), L.P.—Large and roundish-oblong. Skin dark red, and smooth. Of second-rate quality, resembling Wilmot's Early Red. Bush pendulous.

Old Preserver. See *Raspberry.*

Over-All (Bratherton's), L.P.—Large and oblong. Skin red, and hairy. Of second-rate quality. Bush pendulous.

Pastime (Bratherton's), L.P.—Large and roundish. Skin dark red, and hairy. Of second-rate quality. The fruit is often furnished with extra bracts attached to its sides. Bush pendulous.

Perfection (Gregory's), L.P.—Large and roundish. Skin green, and downy. A first-rate variety, and late. Bush pendulous.

Pigeon's Egg. See *Miss Bold.*

Pitmaston Green Gage.—Small and obovate. Skin green, and smooth. A first-rate variety, very sugary, and will hang on the bush till it becomes shrivelled. Bush erect.

Prince of Orange (Bell's), L.P.—Large and oblong. Skin yellow, and downy. Of second-rate quality. Bush pendulous.

Prince Regent (Boardman's), L.P.—Large and roundish. Skin dark red, and smooth. A second-rate variety. Bush spreading.

Princess Royal, L.P.—Large and obovate. Skin greenish-white and hairy. Of first-rate quality. Bush pendulous, and a good bearer.

Profit (Prophet's), L.P.—Large and oblong. Skin green and downy. Of second-rate quality. Bush spreading.

Queen Caroline (Lovart's).—Medium sized, obovate. Skin white and smooth. Of second-rate quality. Bush erect.

Raspberry (*Old Preserver; Nutmeg*).—Fruit small, roundish-oblong. Skin thick, dark red, and hairy. Richly flavoured and sweet. Ripens early. Bush spreading, and a good bearer.

Red Champagne (*Dr. Davies' Upright; Countess of Errol; Ironmonger*, in Scotland).—Small and roundish-oblong, sometimes tapering towards the stalk. Skin rather thick, light red, and hairy. Flavour very rich, vinous, and sweet. Bush very erect, and a good bearer. This is known in Scotland by the name of "Ironmonger.'

Red Mogul.—Small, and roundish-oblong. Skin thin,

red, with a mixture of green, and hairy. Of first-rate quality. Bush spreading, and a good bearer; leaves smooth, by which it is distinguished from Ironmonger.

Red Oval, l.p.—Large and oval. Skin red, and hairy. Of first-rate quality. Bush spreading.

Red Walnut (*Murrey; Eckersley's Double-bearing*).—Medium sized, obovate. Skin red, and downy. An early variety. Of second-rate quality. Bush spreading.

Red Warrington (*Aston; Aston Seedling; Volunteer*).—Above medium size, roundish-oblong. Skin red, and hairy. A first-rate late variety, and highly esteemed for preserving. Bush pendulous.

Rifleman (Leigh's), l.p. (*Alcock's Duke of York; Yates' Royal Anne; Grange's Admirable*). — Large, roundish. Skin red, and hairy. A first-rate late variety. Bush erect, and a good bearer.

Ringleader (Johnson's), l.p. — Large and oblong. Skin red, and smooth. A second-rate variety. Bush pendulous.

Roaring Lion (Farrow's), l.p. (*Great Chance*).—Very large, oblong. Skin red, and smooth. A second-rate variety as to flavour, but one of the largest in size. Bush pendulous.

Rob Roy.—Medium sized, obovate. Skin red, and hairy. A first-rate variety, and very early. Bush erect.

Rockwood (Prophet's), l.p. — Large and roundish. Skin yellow, and hairy. Flavour second-rate. Bush erect.

Rough Red (*Little Red Hairy; Old Scotch Red; Thick-skinned Red*).—Small and round. Skin red, and hairy. A first-rate variety, of excellent flavour, and highly esteemed for preserving. Bush spreading.

Rough Yellow. See *Sulphur*.

Round Yellow. See *Rumbullion*.

Royal White.—Small and round. Skin white, and hairy. A first-rate dessert variety. Bush erect.

Rumbullion (*Yellow Globe; Round Yellow*).—Small and roundish. Skin pale yellow, and downy. Flavour of second-rate quality. Bush erect, and a great bearer; and the fruit much grown for bottling.

RUMBULLION, GREEN.—Small and round. Skin green, and hairy. Flavour second-rate. Bush erect.

SCOTCH NUTMEG.—Medium sized, roundish. Skin red, hairy, or downy. Flavour second-rate. Bush erect.

SHAKESPERE (Denny's), L.P.—Large and roundish. Skin red, and hairy. Of first-rate flavour. Bush erect.

SHEBA QUEEN (Crompton's), L.P.—Large and obovate. Skin white, and downy. Flavour of the first quality. Bush erect. Very similar to Whitesmith.

Sir Sidney Smith. See *Whitesmith* (Woodward's).

Small Dark Rough Red. See *Small Rough Red.*

SMALL RED GLOBE (*Smooth Scotch*).—Small and roundish. Skin smooth, and red. Of first-rate quality, and with a sharp, rich flavour. Bush erect.

SMALL ROUGH RED (*Small Dark Rough Red*).—Small and round. Skin red, and hairy. Of first-rate quality, and early. Bush spreading, and the leaves pubescent.

SMILING BEAUTY (Beaumont's), L.P.—Large and oblong. Skin thin, yellow, and smooth. Of first-rate flavour. Bush pendulous, and a good bearer.

Smooth Amber. See *Amber.*

Smooth Green. See *Green Walnut.*

Smooth Red. See *Turkey Red.*

Smooth Scotch. See *Small Red Globe.*

SNOWBALL (Adams').—Medium sized, roundish. Skin white, and hairy. Of first-rate flavour. Bush pendulous.

SPORTSMAN (Chadwick's), L.P.—Large and obovate. Skin dark red, and smooth. Flavour second-rate. Bush spreading.

SULPHUR (*Rough Yellow*).—Small and roundish. Skin yellow, and hairy. Flavour of first-rate quality. Bush erect, and the leaves not pubescent, by which it is distinguished from Early Sulphur.

Thick-skinned Red. See *Rough Red.*

TOP SAWYER (Capper's), L.P.—Large and roundish. Skin pale red, and hairy. Flavour of second-rate quality. Bush pendulous.

TURKEY RED (*Smooth Red*).—Small and obovate. Skin

smooth, and red. Of first-rate flavour. Bush spreading.

VICTORY (Lomas'), L.P.—Large and roundish. Skin red, and hairy. Of second-rate flavour, but much esteemed for cooking. Bush pendulous.

VICTORY (Mather's), L.P.—Large and obovate. Skin yellow, and smooth. Flavour only second-rate. Bush spreading.

VIPER (Gorton's), L.P.—Large and obovate. Skin greenish yellow, and smooth. Flavour second-rate. Bush pendulous.

Volunteer. See *Red Warrington.*

WELLINGTON'S GLORY, L.P.—Large and roundish-oblong. Skin thin, white, and downy. Flavour of first-rate quality. Bush erect.

WHITE CHAMPAGNE.—Small and roundish-oblong. Skin white, and hairy. Flavour of first-rate quality. Bush erect; leaves pubescent.

WHITE EAGLE (Cook's), L.P.—Large and obovate. Skin white, and smooth. Flavour of first-rate quality. Bush erect.

WHITE FIG.—Small and obovate. Skin white, and smooth. Flavour of first-rate quality, and rich. Bush spreading, but tender.

WHITE LILY.—Medium sized, obovate. Skin white, and downy. Flavour of second-rate quality. Bush erect.

WHITE LION (Cleworth's), L.P.—Large and obovate. Skin white, and downy. Of first-rate quality, and a good late sort. Bush pendulous.

WHITE RASP.—Small and round. Skin white, and smooth. Flavour of second-rate quality. Bush spreading.

WHITESMITH (Woodward's), L.P. (*Whitesmith; Sir Sidney Smith; Hall's Seedling; Lancashire Lass; Grundy's Lady Lilford*).—Large, roundish-oblong. Skin white, and downy. Flavour of first-rate excellence. Bush erect, and a good bearer.

WILMOT'S EARLY RED.—Large and roundish-oblong. Skin dark red, and smooth. Of second-rate quality. Bush pendulous.

WISTASTON HERO (Bratherton's), L.P.—Large and

oblong. Skin green, and hairy. Flavour second-rate. Bush erect.

Yates' Royal Anne. See *Rifleman.*

YAXLEY HERO (Speechley's), L.P.—Large and obovate. Skin red, and hairy. Flavour of first-rate quality. Bush erect.

Yellow Amber. See *Amber.*

YELLOW BALL.—Medium sized, roundish. Skin yellow, and smooth. Flavour of first-rate quality. Bush erect.

YELLOW CHAMPAGNE (*Hairy Amber*). — Small and roundish. Skin yellow, and hairy. Of first-rate excellence. Bush erect.

Yellow Globe. See *Rumbullion.*

YELLOWSMITH. — Small and roundish-oblong. Skin yellow, and hairy. Of first-rate quality, resembling Yellow Champagne. Bush erect.

YELLOW WARRINGTON (*Yellow Aston*).—Middle sized, roundish-oblong. Skin yellow, and hairy. Of first-rate quality. Bush pendulous.

York Seedling. See *Glenton Green.*

---

## SELECT GOOSEBERRIES.

### FOR DESSERT USE.

*Red.*

Ironmonger
Keens' Seedling
Miss Bold
Raspberry
Red Champagne
Red Globe
Red Warrington
Rough Red
Scotch Nutmeg
Small Rough Red
Turkey Red
Wilmot's Early Red

*Yellow.*

Early Sulphur
Glory of Ratcliff
Rockwood
Rumbullion
Yellow Ball
Yellow Champagne

*Green.*

Green Gascoigne
Green Prolific
Green Walnut
Heart of Oak
Hebburn Prolific
Pitmaston Green Gage

*White.*

| | |
|---|---|
| Bright Venus | Hedgehog |
| Crystal | White Champagne |
| Early White | Whitesmith |

---

## FOR EXHIBITION,

### WITH THEIR GREATEST WEIGHTS.

*Red.*

| | Dwts. | Grs. | | Dwts. | Grs. |
|---|---|---|---|---|---|
| Companion......... | 26 | 8 | London........... | 34 | 7 |
| Conquering Hero . | 28 | 3 | Slaughterman ... | 24 | 17 |
| Lion's Provider... | 25 | 8 | Wonderful ...... | 28 | 12 |

*Yellow.*

| | Dwts. | Grs. | | Dwts. | Grs. |
|---|---|---|---|---|---|
| Catherine ......... | 27 | 14 | Leader ........... | 24 | 12 |
| Drill ............... | 27 | 9 | Pilot ............... | 23 | 0 |
| Gunner ........... | 20 | 23 | Railway........... | 22 | 21 |

*Green.*

| | Dwts. | Grs. | | Dwts. | Grs. |
|---|---|---|---|---|---|
| General ............ | 23 | 21 | Queen Victoria... | 23 | 8 |
| Gretna Green...... | 22 | 7 | Thumper ......... | 25 | 6 |
| Over-All........... | 23 | 10 | Turn Out ......... | 22 | 0 |

*White.*

| | Dwts. | Grs. | | Dwts. | Grs. |
|---|---|---|---|---|---|
| Eagle ............... | 20 | 0 | Queen of Trumps | 25 | 12 |
| Freedom........... | 23 | 16 | Snowball ......... | 23 | 8 |
| Lady Leicester ... | 22 | 9 | Snowdrop......... | 23 | 0 |

# GRAPES.

## SYNOPSIS OF GRAPES.

### I. BERRIES ROUND, OR NEARLY SO.

#### * *Black or Purple.*

A. *Muscats.*†

August Muscat
Black Frontignan
Caillaba
Early Black Muscat
July Muscat
Purple Constantia
Sarbelle Muscat

B. *Not Muscats.*

Barbarossa
Bidwill's Seedling
Black Corinth
Black Damascus
Black July
Black Muscadine
Black St. Peter's
Black Sweetwater
Dutch Hamburgh
Esperione
Frankenthal
Miller's Burgundy

#### ** *Red, Tawny, or Striped.*

A. *Muscats.*

Catawba
Red Frontignan
Madeira Muscat

B. *Not Muscats.*

Aleppo
Chasselas de Falloux
Gromier du Cantal
Negropont Chasselas
Red Chasselas

#### *** *White, Yellow, or Green.*

A. *Muscats.*

Chasselas Musqué
Early Saumur Muscat
Muscat Ottonel
White Frontignan

B. *Not Muscats.*

Buckland Sweetwater
Calabrian Raisin
Chaptal
Chasselas Duhamel
Chasselas Vibert
Ciotat
Early Chasselas
Early Malingre
Early White Malvasia
Pitmaston White Cluster
Prolific Sweetwater
Royal Muscadine
White Corinth
White Nice
White Rissling
White Sweetwater

† The term "Muscats" includes, besides the true Muscats, the American Grapes, with their peculiar foxy flavour.

## II. BERRIES OVAL, OR NEARLY SO.

### * *Black or Purple.*

A. *Muscats.*

Black Muscat of Alexandria
Isabella
Muscat Hamburgh

B. *Not Muscats.*

Black Champion
Black Cluster
Black Hamburgh
Black Muscadine
Black Prince
Blussard Noir
Burchardt's Prince
Cambridge Botanic Garden
Gros Maroc
Ischia
Kempsey Alicante
Lady Downe's Seedling
Œillade
Trentham Black
West's St. Peter's

### ** *Red, Tawny, or Striped.*

A. *Muscats.*

None

B. *Not Muscats.*

Lombardy
Morocco
Purple Fontainbleau
Schiras

### *** *White, Yellow, or Green.*

A. *Muscats.*

Bowood Muscat
Charlesworth Tokay
Canon Hall Muscat
Muscat of Alexandria
Muscat St. Laurent

B. *Not Muscats.*

Alexandrian Ciotat
Burchardt's Amber Cluster
Cornichon Blanc
Early Green Madeira
Golden Hamburgh
Marchioness of Hastings
St. John's
Scotch White Cluster
Syrian
Trebbiano
Verdelho
White Lisbon
White Romain
White Tokay

---

Aiga Passera. See *Black Corinth.*

ALEPPO (*Striped Muscadine; Variegated Chasselas; Chasselas Panaché; Morillon Panaché; Raisin d'Alep; Raisin Suisse*).—Bunches medium sized, loose, and not shouldered. Berries medium sized, round, of various colours, some being black, others white or red, while some are striped with black, or red and white; sometimes a bunch will be half white and half black; and others are wholly white or wholly black. The flesh is inferior in flavour. The vine succeeds in a warm vinery, but requires the hothouse to bring it to perfection. The leaves are striped with green, red, and yellow.

ALEXANDRIAN CIOTAT.—Bunches large, long, and loose, with narrow shoulders. Berries oval. Skin thin, pale yellow, but becoming of an amber colour as the fruit are highly ripened, and covered with numerous russety dots. Flesh firm and breaking, juicy, and well flavoured. Ripens with the heat of a vinery. A good bearer, but the bunches set badly.

Alexandrian Frontignan. See *Muscat of Alexandria.*

ALICANTE.—This is a name given to several varieties of grapes in the south of France and in the Peninsula, but is not applicable to any variety in particular. In the department of Gard, it is applied to *Gromier du Cantal.* In Andalusia it is the same as the *Tintilla* and *Tinto* of the same vineyards, the *Mourvéde* of Provence, and *Mataro* of the Eastern Pyrennees. Then the Alicante of Bouches-des-Rhône vineyards is the *Granaxa* of Arragon, and *Granache* of Eastern Pyrennees; while, in the neighbourhood of Alicante, the name is given to two or three different sorts. In Great Britain, *Black Prince* and *Black St. Peter's* are sometimes called Alicante; but a distinct variety from all the above, being sent to me simply under the name of Alicante, I have, to distinguish it, called it *Kempsey Alicante,* which see.

Alicantweine. See *Black Prince.*

Amber Muscadine. See *Royal Muscadine.*

Ansley's Large Oval. See *Morocco.*

Arkansas. See *Catawba.*

D'Arboyce. See *Royal Muscadine.*

AUGUST MUSCAT (*Muscat d'Aoút*).—Berries medium sized, round, inclining to oval. Skin deep purple. Flesh very rich and juicy, with a slight Muscat aroma. An early grape, ripening about the end of August. The vine forms a dwarf bush, and on that account is well adapted for pot culture, but it is a delicate grower. It ripens against a wall.

August Traube. See *Black July.*

Auvergne. See *Black Cluster.*

Auvernat. See *Black Cluster.*

BARBAROSSA (*Brizzola; Rossea; Prince Albert*).—Bunches twelve to eighteen inches long, shouldered, tapering, and compact. Berries round, inclining to oval. Skin tough, but not thick, of a deep black colour, covered with thin bloom. Flesh tender, juicy, and of good flavour, though not rich. A valuable late grape, hanging all the winter; and requires the aid of artificial heat to ripen it. The vine is a bad bearer, except in poor soils.

Barbaroux. See *Gromier du Cantal.*

Bar-sur-Aube. See *Early Chasselas.*

Bec d'Oiseau. See *Cornichon Blanc.*

BIDWILL'S SEEDLING.—This variety, raised at Exeter, has a considerable resemblance to Black Prince, of which it is probably another form. It ripens very well against a wall in the west of England by the end of October.

Black Alicante. See *Black Prince.*

Black Burgundy. See *Black Cluster.*

BLACK CHAMPION (*Champion Hamburgh*).—Bunches with short, thick stalks, not shouldered, thickly set. Berries large, roundish-oval. Skin thin, black, or dark purple, covered with fine thin bloom. Flesh tender, but somewhat firm, very juicy, rich, and sweet; having rarely any stones, or more than one. This is about a fortnight earlier than Black Hamburgh in the same house, and always colours better and more freely than that variety; the berry is also more oval, and the wood shorter jointed. Ripens in a cool vinery.

BLACK CLUSTER (*Auvergne; Auvernat; Black Burgundy; Black Morillon; Burgundy; Blauer Clävner; Early Black; Morillon Noir; Pineau; Schwarzer Riessling*).—Bunches small, very compact, cylindrical, and occasionally shouldered. Berries generally oval, inclining to roundish. Skin thin, blue-black, covered with blue bloom. Flesh juicy, sweet, and richly flavoured. Ripens well against a wall in the open air, and is one of the best for this purpose. The bunches are larger than those of Miller's Burgundy. This is one of the varieties most extensively cultivated for wine on the Rhine and the Moselle, and it also furnishes the greater part of the Champagne and Burgundy wines.

Black Constantia. See *Purple Constantia.*

BLACK CORINTH (*Currant; Corinthe Noir; Passolina*

*Nera; Aiga Passera; Zante*).—Bunches compact, small, and short. Berries small and round, not larger than a pea, with some larger ones interspersed. Skin thin, black, and covered with blue bloom. Flesh juicy, sweet, richly flavoured, and without stones. Requires the heat of a vinery. This variety furnishes the "Currants" of commerce.

BLACK DAMASCUS (*Worksop Manor*).—Bunches large and loose. Berries large and round, interspersed with others of small size. Skin thin, but tough, of a deep black colour. Flesh juicy, sweet, and richly flavoured. A first-rate late grape, requiring the heat of a hothouse to bring it to perfection.

BLACK FRONTIGNAN (*Muscat Noir; Muscat Noir Ordinaire; Sir William Rowley's Black*).—Bunches pretty large, cylindrical, somewhat loose, and occasionally shouldered. Berries small, round, and unequal in size. Skin thin, blue-black, and covered with blue bloom. Flesh firm, red, and juicy, with a rich vinous and musky flavour. Ripens against a wall in favourable situations and in warm seasons; but is generally grown in a vinery.

BLACK HAMBURGH (*Hampton Court; Knevett's Black Hamburgh; Red Hamburgh; Warner's Hamburgh; Blauer Trollinger; Maroquin d'Espagne*).—Bunches large, broadly shouldered, conical, and well set. Berries roundish-oval. Skin thin, but membranous, deep blue-black, covered with blue bloom. Flesh rather firm, but tender, very juicy, rich, sugary, and highly flavoured.

This highly-popular grape succeeds under every form of vine culture. It ripens against a wall, in favourable situations, in the open air. It succeeds well in a cool vinery; and it is equally well adapted for forcing. The vine is a free bearer; and the fruit will hang, under good management, till January and February.

BLACK JULY (*Early Black July; July; Madeleine; Madeleine Noir; Morillon Hâtif; Raisin Précoce; De St. Jean; August Traube; Jacob's Traube*).—Bunches small and cylindrical. Berries small and round. Skin thick, deep purple, covered with blue bloom. Flesh sweet and juicy, but not highly flavoured. Its chief recommendation is its great earliness, and the facility with which it ripens against a wall in the open air. The flowers are tender, and, consequently, unless grown in a cool vinery, the bunches are loose, and the berries thin; but when protected, the plant produces close, compact bunches, and is an excellent bearer. Although this is the earliest grape, it is not so highly flavoured as Black Cluster and Miller's Burgundy.

Black Lisbon. See *Black Prince.*

Black Lombardy. See *West's St. Peter's.*

Black Morillon. See *Black Cluster.*

Black Morocco. See *Morocco.*

Black Muscadel. See *Morocco.*

Black Muscadine (*Black Chasselas; Chasselas Noir*).—Bunches medium sized, compact. Berries about medium sized, round, inclining to oval. Skin thick, deep purplish-black, covered with blue bloom. Flesh juicy, sweet, sugary, and richly flavoured. When well ripened, this is an excellent grape, and has a trace of musky aroma in its flavour; but, to obtain it thus, it requires to be grown in a warm vinery.

Black Muscat of Alexandria (*Red Muscat of Alexandria*).—Bunches large and shouldered. Berries large and oval. Skin thick, dark reddish-purple. Flesh firm and crackling, with a rich, sugary, and musky flavour. A first-rate grape. The berries are rather smaller than those of the White Muscat of Alexandria, but are equally rich in flavour, and ripen more easily. It may be grown either in a warm vinery, or a hothouse; but the latter is not indispensable.

Black Palestine. See *Black St. Peter's.*

Black Portugal. See *Black Prince.*

Black Prince (*Alicante; Boston; Pocock's Damascus; Sir A. Pytche's Black; Steward's Black Prince; Blauer von Alicant; Alicantenwein*).—Bunches long, and generally without shoulders; but occasionally shouldered. Berries above medium size, oval. Skin thick, deep purplish-black, covered with thick blue bloom. Flesh white, or greenish, tender, very juicy, with a rich, sugary, and sprightly flavour. The seed-bearing string (placenta), which is drawn out when the berry is separated from the stalk, has a crimson streak in it. This is a grape of first-rate quality, ripens well in a cool vinery, or against a wall, in favourable situations; and always colours well. The vine is a good bearer; the leaves in autumn die off, beautifully variegated with red, green, and yellow.

This is the *Alicant* and *Black Spanish* of Speechly, and, according to him, it is also called *Lombardy;* but the true Black Spanish is *Black St. Peter's,* and it is sometimes called *Alicante.* It is also the *Blauer von Alicante* of Fintlemann, and the *Alicantweine* of Christ.

Black St. Peter's (*Alicante; Black Lisbon; Black Portugal; Black Palestine; Black Spanish; Black Valentia; St. Peter's; Espagne Noir; Sanct Peters Traube; Schwarzer Spanischer*).—Bunches large and long, sometimes shouldered. Berries above medium size, round. Skin thin, deep blue-black, and covered with bloom. Flesh tender, juicy, and with a rich, brisk flavour. An excellent late grape that will hang till March. It requires to be grown in a warm vinery; but will not bear much forcing, otherwise the berries are liable to crack.

Blacksmith's White Cluster. See *Scotch White Cluster.*

Black Spanish. See *Black Prince.*

Black Spanish. See *Black St. Peter's.*

Black Sweetwater (*Waterzoet Noir*).—Bunches small, short, and compact. Berries round. Skin very thin, and black. Flesh tender, juicy, and very sweet; but has little aroma or richness. This succeeds well against a wall, where it ripens early, or in a cool vinery; but it is impatient of forcing, and the berries are liable to crack when subjected to too much heat.

Black Tripoli.—The Black Tripoli grown at Welbeck since the time of Speechly has long been considered a distinct variety. By some it has been stated to be identical with the Black Hamburgh, and others have as distinctly asserted that it is totally different from that variety. When it is considered that there are two varieties of grapes cultivated in the country under the name of Black Hamburgh, this diversity of opinion is easily accounted for. From the true Black Hamburgh it is certainly distinct; but with the *Frankenthal,* which is also grown under that name, it is as certainly identical. See *Frankenthal.*

Black Valentia. See *Black St. Peter's.*

Blanc Précoce de Kienzheim. See *Early Kienzheim.*

Blanche. See *St. John's.*

Blauer von Alicant. See *Black Prince.*

Blauer Clävner. See *Black Cluster.*

Blauer Müllerrebe. See *Miller's Burgundy.*

Blauer Trollinger. See *Black Hamburgh.*

Blue Frontignan. See *Purple Constantia.*

BLUSSARD NOIR.—Bunches small and rather loose, not shouldered. Berries medium sized, roundish-oval. Skin rather thin, black, and covered with bloom. Flesh tender, juicy, sweet, and richly flavoured. The vine is a very strong grower, but a bad bearer. It is earlier than Black Hamburgh.

Boston. See *Black Prince.*

Boudalès. See *Œillade.*

BOWOOD MUSCAT.—This is a seedling raised from Muscat of Alexandria, to which it bears a close resemblance, but it differs from its parent in setting its fruit better, and in being a better bearer, and much earlier. It is an excellent grape.

Brizzola. See *Barbarossa.*

BUCKLAND SWEETWATER.—Bunches large, shouldered, and well set, heart-shaped. Berries large, round, inclining to oval. Skin thin, transparent, pale green, becoming pale amber when ripe. Flesh tender, melting, and very juicy, sweet, and well flavoured. Seeds rarely more than one in each berry. It ripens in a cool vinery.

BURCHARDT'S AMBER CLUSTER. — Bunches medium sized, conical. Berries medium sized, oval. Skin thin, yellowish-white, becoming amber coloured when ripe. Flesh very juicy, rich, and sugary. Earlier than the Royal Muscadine, and a first-rate grape.

BURCHARDT'S PRINCE.—Bunches long and tapering, larger than those of the Black Prince. Berries medium sized, roundish-oval. Skin thick, of a deep black colour, covered with dense bloom. Flesh firm, juicy, melting, rich, and vinous. An excellent late grape, requiring heat.

Burgundy. See *Black Cluster.*

Busby's Golden Hamburgh. See *Golden Hamburgh.*

CAILLABA (*Caillaba Noir Musquée*).—Bunches long. Berries rather below medium size, round. Skin thin, but membranous, black. Flesh tender, juicy, and sweet, with a Muscat flavour. This is a moderately early grape, and ripens in a cool vinery about the beginning or middle of September. The vine is delicate, and requires high cultivation.

CALABRIAN RAISIN (*Raisin de Calabre*). — Bunches large, slightly shouldered, long, and tapering, sometimes upwards of a foot in length. Berries large, quite round. Skin thick, but so transparent that the texture of the flesh and the stones are distinctly visible; white. Flesh moderately firm, with a sugary juice and good flavour. This is a late and long-hanging grape, forming an excellent white companion to Black St. Peter's. It is not of first-rate quality as to flavour; but is, nevertheless, a valuable grape to grow on account of its late-keeping properties. The vine is a strong grower and a good bearer; succeeds in a cool vinery, and will also stand a good deal of heat.

CAMBRIDGE BOTANIC GARDEN.—This has been said to be identical with Black Prince, with which it has now, in many instances, got confounded; but it differs from that variety in having shorter and much more compact bunches. Bunches rarely shouldered. Berries large and oval. Skin brownish-black. Flesh firm, juicy, sweet, and highly flavoured; with from two to three stones in each berry: while in Black Prince they vary from three to five.

An excellent out-door grape, ripening well against a wall, and well adapted for a cold vinery. Mr. Rivers has found it well suited for pot culture.

Campanella Bianca. See *Royal Muscadine.*

CANON HALL MUSCAT.—This differs from its parent, the Muscat of Alexandria, in having better-set and more tapering bunches, and rather larger and longer berries. The vine is of more robust growth, and the flowers have six and sometimes seven, stamens; but the fruit is not so highly flavoured as Muscat of Alexandria.

CATAWBA (*Arkansas; Catawba Tokay; Lebanon Seedling; Red Murrey; Singleton*).—Bunches medium sized, shouldered. Berries medium sized, round. Skin thick, pale red, becoming a deeper colour as it ripens, and covered with a lilac bloom. Flesh somewhat glutinous, juicy, sweet, and musky. A popular American dessert grape, and used also for wine. It is very productive, and very hardy.

Champion Hamburgh. See *Black Champion.*

Chaptal.—Bunches large. Berries large and round, inclining to oval. Skin white. Flesh juicy and sweet. This is a new French grape of excellent quality, well adapted for a cool vinery, when it ripens about the middle of September. The vine is a great bearer, and, according to Mr. Rivers, is well adapted for pot culture.

Charlesworth Tokay.—This is very much like Muscat of Alexandria. Some consider it quite distinct, but I have as yet failed to observe wherein it differs. If it is distinct, it is not sufficiently so to make two varieties of them.

Chasselas. See *Royal Muscadine.*

Chasselas Blanc. See *Royal Muscadine.*

Chasselas Bleu de Windsor. See *Esperione*

Chasselas Dorée. See *Royal Muscadine.*

Chasselas Duhamel.—This is, in all respects, very much like Chasselas Vibert, and was raised in the same batch of seedlings. Mr. Rivers describes it to me as a fine, large, amber-coloured Sweetwater-like sort, which is likely to prove very valuable. He imported it for the first time into this country three or four years ago.

Chasselas de Falloux (*Chasselas Rose de Falloux*).—Bunches long and compact. Berries large, round, and somewhat flattened. Skin tough, of a pale yellow colour at first, but gradually changing to a pale red. Flesh firm, juicy, sweet, and refreshing, with a distinct musky flavour. The vine is a great bearer, and well suited for pot culture. The fruit ripens in September in an ordinary vinery.

Chasselas de Fontainbleau. See *Royal Muscadine.*

Chasselas Musqué (*Josling's St. Albans; Muscat Fleur d'Orange; Muscat de Jesus; Muscat Primavis; Pascal Musqué; Tokai Musqué*).—Bunches long, tapering, rather loose, and shouldered. Berries above medium size, round. Skin greenish-white, changing to pale amber when highly ripened, and covered with a delicate white bloom. Flesh firm, rich, sugary, and with a high Muscat flavour. A most delicious grape of first-rate quality. It may be grown either in a cool or warm vinery; but the berries are very liable to crack, unless the vine is growing in a shallow border, and the roots are kept moderately dry when the fruit is ripening. It is rather an early variety, and ripens in a vinery in the beginning of September.

Chasselas de Negrepont. See *Negropont Chasselas.*

Chasselas Panaché. See *Aleppo.*

Chasselas Rose de Falloux. See *Chasselas de Falloux.*

CHASSELAS VIBERT.—Bunches long and loose. Berries large and round. Skin thin and transparent, yellowish-white, but when highly ripened of a fine pale amber colour. Flesh tender, juicy, and sweet. This, in the form and size of the bunches and berries, resembles the Prolific Sweetwater; but it is readily distinguished from all the Sweetwaters, to which section it belongs, by the bristly pubescence of its leaves, both above and beneath. Mr. Rivers informs me, that it ripens with him ten or twelve days before the Royal Muscadine; that the vine is hardy and prolific, and well adapted for pot culture. It may be grown in a cool vinery.

Cinq Saous. See *Œillade.*

CIOTAT (*Parsley-leaved; Raisin d'Autriche; Petersilien Gutedel*).—Bunches medium sized, not quite so large as those of Royal Muscadine, shouldered and loose. Berries medium sized, round, uneven, with short, thin stalks. Skin thin, greenish-yellow or white, covered with bloom. Flesh tender, sweet, and with the flavour of Royal Muscadine, of which this variety is a mere form, differing in having the leaves very much cut. It ripens in a cool vinery.

Le Cœur. See *Morocco.*

Corinthe Blanc. See *White Corinth.*

Corinthe Noir. See *Black Corinth.*

CORNICHON BLANC (*Finger Grape; White Cucumber; Bec d'Oiseau; Teta de Vaca*).—Bunches rather small, round, and loose. Berries very long, sometimes an inch and a half, and narrow; tapering to both ends, and just like very large barberries. Skin thick, green, and covered with white bloom. Flesh firm and sweet. A late-ripening and late-hanging grape of little value, and requires stove heat to ripen it.

Cumberland Lodge. See *Esperione.*

Currant. See *Black Corinth.*

De Candolle. See *Gromier du Cantal.*

DUTCH HAMBURGH (*Wilmot's Hamburgh*).—Bunches medium sized, compact, and rarely shouldered. Berries very large, roundish-oblate, uneven and hammered. Skin thick, very black, and covered with a thin bloom. Flesh pretty firm, coarse, and not so highly flavoured as the Black Hamburgh. It ripens in an ordinary vinery.

Dutch Sweetwater. See *White Sweetwater.*

Early Black. See *Black Cluster.*

EARLY BLACK MUSCAT (*Muscat Précoce d'Août.*—Mr. Rivers' description of this variety, which I have not seen, is—Berries below medium size, and round. Skin black. Flesh rich and juicy, with a rich Frontignan flavour. The vine is more robust in its habit than the August Muscat, and the fruit ripens against a wall. This is one of the seedlings of the late M. Vibert, of Angers.

EARLY CHASSELAS (*Chasselas Hâtif; Bar-sur-Aube; Krach Gutedel*).—This is very similar to the Royal Muscadine in general appearance, and has, therefore, been frequently confounded with it; but it is a very distinct variety when obtained true, and is readily known by its very firm crackling flesh, which is richly flavoured. The vine may be distinguished by its small quantity of foliage, which is somewhat hairy, and by the leaftstalk being frequently warted.

Early Leipzic. See *Early White Malvasia.*

EARLY KIENZHEIM (*Blanc Précoce de Kienzheim; Précoce de Kienzheim*).—Bunches small, cylindrical and well set. Berries about medium size, roundish-oval or oval. Skin tender, white, and transparent, covered with a very thin bloom. Flesh very tender and juicy, sweet and pleasantly flavoured, like the Sweetwater.

This is one of the earliest grapes known, and ripens in a cool vinery from the beginning to the middle of August. It will also succeed against a wall in the open air; but, of course, is not then so early.

EARLY MALINGRE (*Malingre; Précoce de Malingre;*

*Précoce Blanc*).—Bunches of pretty good size. Berries round, inclining to oval, and of medium size. Skin white. Flesh rather richly flavoured, juicy and sugary. One of the earliest grapes, ripening in a cool vinery in the beginning of August; and, in the open air, against a wall, it is the earliest white grape. The vine is a most abundant bearer, forms a handsome bush, and is well suited for pot culture.

EARLY SAUMUR MUSCAT (*Muscat de Saumur; Madeleine Musqué de Courtiller; Précoce Musqué*).—Bunches rather compact. Berries medium sized and round. Skin white, assuming an amber tinge towards maturity. Flesh firm and crackling, rich and sugary, with a distinct, but not strong, Muscat flavour. This is one of the earliest grapes, ripening with the Black July, from seed of which it was raised.

The vine is an abundant bearer. It is an excellent grape, and may be grown either in a cool vinery, or against a wall in the open air.

EARLY WHITE MALVASIA (*Grove-End Sweetwater; Early Leipzic; Morna Chasselas; White Melier; Melier Blanc Hâtif; Früher Leipziger; Weisse Cibebe*).—Bunches rather large, six to eight inches long, loose, tapering, and occasionally shouldered. Berries large, round, inclining to oval. Skin thin and transparent, greenish-white, but becoming yellow at maturity, and covered with white bloom. Flesh abundant, very juicy, sweet, and with a rich flavour. Ripens in a cool vinery about the end of August, and also against a wall in the open air.

The vine is an excellent bearer, and succeeds well when grown in pots.

ESPERIONE (*Cumberland Lodge; Turner's Black; Aspirant Noir; Espiran; Chasselas Bleu de Windsor*).—Bunches large and shouldered. Berries large, round, and inclining to oblate. Skin dark blackish-purple, covered with blue bloom. Flesh rather firm than tender, juicy, sweet, and well flavoured; but inferior to the Black Hamburgh.

This is a variety bearing a close resemblance to the Frankenthal. Its great recommendation is its ripening so well out of doors against a wall, for which it is said to be better adapted, and where it ripens better than the Black Hamburgh, and ten or fifteen days earlier. It is distinguished from Black Hamburgh by its leaves dying off a rich purple colour, and not yellow.

Finger Grape. See *Cornichon Blanc.*

Flame-coloured Tokay. See *Lombardy.*

Frankenthal. See page 121.

Froc de la Boulaye. See *Prolific Sweetwater.*

Früher Leipziger. See *Early White Malvasia.*

GOLDEN HAMBURGH (*Busby's Golden Hamburgh; Stockwood Park Golden Hamburgh*).—Bunches large, loose, branching, and shouldered. Berries large and oval. Skin thin, of a pale yellow colour; but when highly ripened, pale amber. Flesh tender and melting, very juicy, rich, sugary, and vinous. An excellent grape. Ripens in a cool vinery, and forces well.

Grec Rouge. See *Gromier du Cantal.*

GROMIER DU CANTAL (*Barbaroux; De Candolle; Grec Rouge; Gros Gromier du Cantal; Malaga; Raisin du Pauvre; Raisin de Servie*).—Bunches large, a foot long, broad, and shouldered. Berries large and round. Skin very thin, amber coloured, mottled with light purplish-brown. Flesh tender, juicy, and sweet, with a brisk vinous flavour. Requires a warm vinery to ripen it; and it does not keep long after being ripe. In some of the vineyards of France, and particularly in those of Tarn-et-Garonne, it is called *Alicante.*

Gros Coûlard. See *Prolific Sweetwater.*

Gros Gromier du Cantal. See *Gromier du Cantal.*

GROS MAROC (*Marocain*).—Bunches large, long, and shouldered, and with a long stalk. Berries large and oval. Skin thick, of a deep reddish-purple, and covered with an abundant blue bloom. Flesh tender, sweet, and richly flavoured.

This is an excellent grape, and ripens along with the Black Hamburgh. It is frequently confounded with the Gros Damas, from which it is distinguished by its smoother and more deeply-cut leaves, shorter-jointed wood, and earlier ripening.

GROS ROMAIN.—This is a variety introduced by Mr. Rivers, of which he speaks very highly. I have never yet seen the fruit; but he informs me that it is a most delicious grape, with very large, round, amber-coloured berries, almost yellow. It is quite a distinct variety from White Romain, or, as it is sometimes called, Muscat Romain.

Grove-End Sweetwater. See *Early White Malvasia.*

Gutedel. See *Royal Muscadine.*

Hampton Court. See *Black Hamburgh.*

Horsforth Seedling. See *Morocco.*

Isabella.—Bunches large. Berries large and oval. Skin thin, of a dark purple colour, almost black, and covered with bloom. Flesh tender, juicy, sweet, and vinous, with a musky flavour.

This is a variety of Vitis Labrusca, a native American grape, cultivated in the open air in the United States, both for the dessert and for wine; but it is not of much account in England.

Ischia (*Noir Précoce de Gênes; Uva di tri volte*).—The bunches and berries of this variety very much resemble those of Black Cluster; but the fruit ripens as early as that of Black July, and is very much superior in flavour to that variety. Berries medium sized, black, very juicy, sweet, and vinous. The vine is very vigorous and luxuriant in its growth, and bears abundantly, if not pruned too close. In Italy it produces three crops in a year by stopping the shoot two or three joints beyond the last bunch just as the flower has fallen and the berries set; new shoots are started from the joints that are left, and also bear fruit, and these being again stopped, a third crop is obtained.

This variety succeeds admirably against a wall in the open air.

Jacob's Traube. See *Black July.*

Jew's. See *Syrian.*

Joannec. See *St. John's.*

Joannenc. See *St. John's.*

Josling's St. Alban's. See *Chasselas Musqué.*

July. See *Black July.*

July Muscat (*Muscat de Juillet*).—This is a very early variety, having a distinct Muscat flavour. It was introduced by Mr. Rivers, who describes it as follows:—Berries round, purple; of medium size; rich, juicy, and excellent. This grape will ripen on a wall, as it is one of the earliest of its race, and is well adapted for pot culture in the orchard-house.

KEMPSEY ALICANTE.—Bunches six to eight inches long, not shouldered, and rather thickly set. Berries very large, from an inch to an inch and a quarter long, and three quarters to an inch wide; oval. Skin thick and tough, of a deep blue-black colour at the apex when ripe, but towards the stalk of a greenish-yellow, mottled with dark purple. Flesh greenish, firm, sweet, and with a fine aroma when fully ripe. Seeds generally one or two only, but sometimes four.

The berries, in size and colour, are more like plums. The vine is a free grower, a good bearer, and requires a high temperature to ripen the fruit thoroughly. The foliage when young is very thin and tender, and covered with a delicate down. This is a very late grape, being fully three weeks or a month later than any other variety; still it forces well, and may also be grown in pots. It will hang till May.

Knevett's Black Hamburgh. See *Black Hamburgh.*

Krach Gutedel. See *Early Chasselas.*

Laan Hâtif. See *Scotch White Cluster.*

LADY DOWNE'S SEEDLING.—Bunches shouldered, eight to ten inches long, and rather loose. Berries above medium size, ten-twelfths of an inch long and nine-twelfths wide; oval. Skin rather thick, tough, and membraneous, reddish-purple at first, but becoming quite black when fully coloured, and covered with a delicate bloom. Flesh dull opaline white, firm, sweet, and richly flavoured, with a faint trace of Muscat flavour, but not so much as to include it among Muscats. Seeds generally in pairs.

This is a very valuable grape, and may be ripened with the heat of an ordinary vinery. It forces well, and will hang till the month of March without shrivelling or discoloration of either berries or stalks. The vine is a vigorous grower and an abundant bearer, seldom producing less than three bunches on each shoot. I have seen bunches of this grape ripened in August, hang till March, and preserve all their freshness even at that late season, when the berries were plump and delicious.

Lashmar's Seedling. See *St. John's.*

Lebanon Seedling. See *Catawba.*

Lombardy (*Flame-coloured Tokay; Red Rhenish; Red Taurida; Wantage*).—Bunches very large, shouldered, closely set, and handsome; sometimes weighing from six to seven pounds. Berries large and round, inclining to oval. Skin pale red or flame coloured. Flesh firm, sweet, and well flavoured, but only second-rate.

This requires a high temperature to ripen it. The vine is a very strong grower, and requires a great deal of room; but it is a good bearer. The only recommendation to this variety is the great size of the bunches and beauty of the fruit.

Macready's Early White. See *St. John's.*

Madeira Muscat.—Bunches of medium size, rather compact. Berries above medium size, round. Skin reddish-purple. Flesh very juicy and rich, with a high musky flavour.

This is an excellent grape, and ripens well in a cool vinery at the same season as the Black Hamburgh.

Madeleine. See *Black July.*

Madeleine Blanche de Malingre. See *Early Malingre.*

Madeleine Musqué de Courtiller. See *Early Saumur Muscat.*

Madeleine Noir. See *Black July.*

Malaga. See *Gromier du Cantal.*

Malaga. See *Muscat of Alexandria.*

Malingre. See *Early Malingre.*

Marchioness of Hastings.—Bunches large, loose, and broadly shouldered. Berries upwards of an inch long and about an inch wide; oval. Skin thin, greenish white, covered with thin grey bloom. Flesh squashy and watery, without much flavour. This is an early grape, and ripens in an ordinary vinery. Its only recommendation is the size of the bunches, which may be grown to weigh four pounds.

Maroquin d'Espagne. See *Black Hamburgh.*

Melier Blanc Hâtive. See *Early White Malvasia.*

Merrick's Victoria. See *Frankenthal.*

Meunier. See *Miller's Burgundy.*

Mill Hill Hamburgh.—This is so much like *Dutch Hamburgh* as not to be distinguishable from it, but there are some who maintain that they are distinct. The only difference I have ever been able to detect is, when the foliage is young that of Mill Hill appears paler and more waved than that of Dutch Hamburgh, but that character soon disappears, and the two become to all appearance the same.

Miller Grape. See *Miller's Burgundy.*

MILLER'S BURGUNDY (*Miller Grape; Meunier; Blauer Müllerrebe*).—Bunches short, cylindrical, and compact, with a long stalk. Berries small, round, inclining to oval, uniform in size, with short-warted stalks. Skin thin, black, and covered with blue bloom. Flesh red, sweet, juicy, and highly flavoured, and contains two seeds.

An excellent grape for out-door cultivation, as it ripens well against a wall. It is easily distinguished from all other grapes by its very downy leaves, which, when they are first expanded, are almost white, and this they in some degree maintain during the greater part of the season. On this account it is called "The Miller."

Mogul. See *Morocco.*

Money's St. Peter's. See *West's St. Peter's.*

Morillon Hâtif. See *Black July.*

Morillon Noir. See *Black Cluster.*

Morillon Panaché. See *Aleppo.*

MOROCCO (*Ansley's Large Oval; Black Morocco; Black Muscadel; Le Cœur; Horsforth Seedling; Mogul; Red Muscadel*).—Bunches large and shouldered. Berries of unequal size; some are large and oval. Skin thick, reddish-brown, becoming blackish-brown when fully ripe; beginning to colour at the apex and proceeding gradually towards the stalk, where it is generally paler. Flesh firm, sweet, but not highly flavoured; the small berries are generally without stones, and the large ones have rarely more than one.

This is only a second-rate grape as regards flavour. It is very late, and requires stove heat to ripen it thoroughly.

Moscatel Commun. See *White Frontignan.*

Moscatel Menudo. See *Red Frontignan.*

MUSCAT OF ALEXANDRIA (*Alexandrian Frontignan; Malaga; Muscat of Jerusalem; Muscat of Lunel; Panse Musqué; Passe Musqué; Tottenham Park Muscat*).—

Bunches large, long, loose, and shouldered; stalk long. Berries large, oval, unequal in size, and with long, slender, warted stalks. Skin thick, generally greenish yellow; but, when highly ripened, a fine pale amber colour, and covered with thin white bloom. Flesh firm and breaking, not very juicy, but exceedingly sweet and rich, with a fine Muscat flavour.

A well-known and most delicious grape, requiring a high temperature to ripen it thoroughly; but it may be sufficiently ripened in a warm vinery, provided it has a high temperature at the time of flowering and while the fruit is setting. The vine is an abundant bearer, but the bunches set badly. To remedy this defect, a very good plan is to draw the hand down the bunches when they are in bloom so as to distribute the pollen, and thereby aid fertilisation.

It is this grape which furnishes the Muscatel Raisins, imported in boxes from Spain.

Muscat d'Aout. See *August Muscat.*

Muscat Blanc. See *White Frontignan.*

Muscat Fleur d'Orange. See *Chasselas Musqué.*

Muscat Gris. See *Red Frontignan.*

MUSCAT HAMBURGH (*Snow's Muscat Hamburgh*).—Bunches above medium size, compact, and shouldered. Berries rather large, varying from round to oval. Skin tough, but not thick, deep purplish-black, covered with thin blue bloom. Flesh tender, very juicy, rich, and sugary, with a fine Muscat aroma.

This excellent grape may be ripened in a house subjected to the same amount of heat as is generally given to the Black Hamburgh, and it has also been ripened in a cool vinery.

Muscat of Jerusalem. See *Muscat of Alexandria.*

Muscat de Jesus. See *Chasselas Musqué.*

Muscat de Juillet. See *July Muscat.*

MUSCAT ST. LAURENT.—Bunches similar to those of Royal Muscadine. Berries small, roundish oval. Skin thin, greenish-yellow, becoming pale amber when thoroughly ripened. Flesh very tender, melting, and juicy, with a refreshing, juicy, and a distinct Muscat aroma. This variety, introduced by Mr. Rivers, that gentleman says, is very early, and will ripen on a wall with the

Sweetwater. It is well adapted for pot culture in the orchard-house, and in cool vineries.

Muscat of Lunel. See *Muscat of Alexandria.*

Muscat de Naples. See *Purple Constantia.*

Muscat Noir. See *Black Frontignan.*

Muscat Noir d'Espagne. See *Trentham Black.*

Muscat Noir Ordinaire. See *Black Frontignan.*

MUSCAT OTTONEL.—This is an early variety of Muscat grape, introduced by Mr. Rivers, of Sawbridgeworth, and which ripens its fruit in a cool vinery. That gentleman describes it as having a compact bunch, and round, white, and rather small berries. "A very hardy, nice Muscat grape."

Muscat Précoce d'Août. See *Early Black Muscat.*

Muscat Primavis. See *Chasselas Musqué.*

Muscat Romain. See *White Romain.*

Muscat Rouge. See *Red Frontignan.*

Muscat de Sarbelle. See *Sarbelle Muscat.*

Muscat de Saumur. See *Early Saumur Muscat.*

NEGROPONT CHASSELAS (*Chasselas de Negrepont*).—This is a variety which, in the bunches and foliage, resembles the Royal Muscadine. The berries are at first of a pale green colour, and gradually become of a fine clear red as they attain maturity. In this respect they differ from those of the Red Chasselas, which, from their setting, are of a bright red colour. Like the Royal Muscadine, it is of excellent flavour, and early.

Nepean's Constantia. See *White Frontignan.*

Noir Précoce de Gênes. See *Ischia.*

ŒILLADE (*Ulliade; Boudalès; Cinq Saous; Prunelas*).—Bunches medium sized, and with long stalks. Berries large, oval, uniform in size, and dangling from long stalks. Skin thin, of a dark purplish-black colour, and covered with bloom. Flesh rather firm, and breaking, juicy, sweet, and of good flavour.

The vine is a very abundant bearer, and ripens its fruit in a cool vinery.

Oldaker's St. Peter's. See *West's St. Peter's.*

Palestine. See *Syrian.*

Panse Musqué. See *Muscat of Alexandria.*

Parsley-leaved. See *Ciotat.*

Passe Musqué. See *Muscat of Alexandria.*

Passolina Nera. See *Black Corinth.*

Perle Blanche. See *White Sweetwater.*

Petersilien Gutedel. See *Ciotat.*

Pineau. See. *Black Cluster.*

Pitmaston White Cluster.—Bunches medium sized, compact, and shouldered. Berries medium sized, round, inclining to oblate. Skin thin, amber coloured, and frequently russety. Flesh tender and juicy, sweet and well flavoured. An excellent early grape; succeeds well in a cool vinery, and ripens against a wall in the open air.

Pocock's Damascus. See *Black Prince.*

Poonah. See *West's St. Peter's.*

Pope Hamburgh. See *Frankenthal.*

Précoce Blanc. See *Early Malingre.*

Précoce de Kienzheim. See *Early Kienzheim.*

Précoce de Malingre. See *Early Malingre.*

Précoce Musqué. See *Early Saumur Muscat.*

Prince Albert. See *Barbarossa.*

Prolific Sweetwater (*Froc de la Boulaye; Gros Coulard*).—Bunches medium sized, cylindrical, loose, and not shouldered. Berries large and round, uniform in size. Skin thin, greenish-yellow, but pale amber when fully ripe. Flesh tender, juicy, and sweet, with an excellent flavour.

This is an excellent early white grape, and sets its fruit much better than the old Sweetwater. It ripens well in a cool vinery, and is well adapted for pot culture.

Prunelas. See *Œillade.*

Purple Constantia (*Black Constantia; Purple Frontignan; Blue Frontignan; Violet Frontignan; Muscat de Naples; Violette Muskateller*).—Bunches long and tapering, very much more so than those of Black Frontignan, and with small shoulders. Berries large and round. Skin dark purple, covered with thick blue bloom. Flesh juicy, very richly flavoured, and with a Muscat aroma which is less powerful than in Black Frontignan.

This is a most delicious grape, and requires to be grown

in a warm vinery. It is the *Black* or *Purple Frontignac* of Speechly; but is very different from what is generally cultivated for Black Frontignan—that variety being the Blue or Violet Frontignac of Speechly.

PURPLE FONTAINBLEAU.—I have never seen this grape, but Mr. Rivers speaks of it as a very hardy variety, ripening against a wall in the open air; well adapted for pot culture, and a prodigious bearer. The berries are oval, light purple, sweet, and juicy.

Purple Frontignan. See *Purple Constantia.*

Raisin d'Alep. See *Aleppo.*

Raisin d'Autriche. See *Ciotat.*

Raisin de Calabre. See *Calabrian Raisin.*

Raisin des Carmes. See *West's St. Peter's.*

Raisin de Cuba. See *West's St. Peter's.*

Raisin de Frontignan. See *White Frontignan.*

Raisin du Pauvre. See *Gromier du Cantal.*

Raisin Précoce. See *Black July.*

Raisin de St. Jean. See *St. John's.*

Raisin de Servie. See *Gromier du Cantal.*

Raisin Suisse. See *Aleppo.*

RED CHASSELAS (*Red Muscadine; Chasselas Rouge; Chasselas Rouge Foncé; Cerese; Septembro*).—Bunches medium sized, loose, rarely compact, shouldered; with long, thin, and somewhat reddish stalks. Berries medium sized, round. Skin thin, red, covered with a violet bloom. Flesh juicy and sweet. The vine is a great bearer, and will ripen its fruit in a cool vinery. The most remarkable character of this variety is, that from the time the germ is visible, or, as Mr. Rivers says, "no bigger than a pin's head, it changes to red," and it becomes gradually paler as the fruit ripens. Mr. Rivers says, "it is as good as Royal Muscadine when fully ripe, and a great bearer."

RED FRONTIGNAN (*Grizzly Frontignan; Muscat Gris; Muscat Rouge; Moscatel Menudo; Cevana Dinka; Rother Muskateller; Grauer Muskataller*).—Bunches large, long, and generally cylindrical, but occasionally with very small shoulders. Berries above medium size, round. Skin rather thick, yellow on the shaded side, clouded with pale red on the side next the sun, and covered with grey bloom. Flesh rather firm, juicy, but not very melting, with a rich, sugary, and musky flavour.

Ripens about the end of September when not forced, and requires the heat of a warm vinery.

Red Hamburgh. See *Black Hamburgh.*

Red Muncy. See *Catawba.*

Red Muscadel. See *Morocco.*

Red Muscadine. See *Red Chasselas.*

Red Muscat of Alexandria. See *Black Muscat of Alexandria.*

Red Rhenish. See *Lombardy.*

Red Taurida. See *Lombardy.*

Rheingauer. See *White Rissling.*

Riessling. See *White Rissling.*

Rösslinger. See *White Rissling.*

Rossea. See *Barbarossa.*

Rother Muskateller. See *Red Frontignan.*

ROYAL MUSCADINE (*Amber Muscadine; Muscadine; White Chasselas; D'Arboyce; Chasselas; Chasselas Doré; Chasselas de Fontainbleau; Campanella Bianca; Weisser Gutedel*).—Bunches long, loose, and shouldered; sometimes compact and cylindrical. Berries large, round, and, in the compact bunches, inclining to oval. Skin thin and transparent, greenish-yellow, becoming pale amber when quite ripe, and sometimes marked with tracings and dots of russet; covered with thin white bloom. Flesh tender and juicy, sweet, and richly flavoured.

This excellent and well-known grape ripens well in a cool vinery and against walls in the open air. The many names it has received have arisen from the various forms it frequently assumes, and which are occasioned entirely by the nature of the soil and the different modes of treatment to which it is subjected. There is no real difference between this, the common Chasselas, and Chasselas de Fontainbleau. The White Muscadine of some authors is the Early Chasselas.

Rüdesheimer. See *White Rissling.*

St. Jean. See *Black July.*

ST. JOHN'S (*Raisin de St. Jean; Joannec; Joannenc;*

*Blanche; Lashmar's Seedling; Macready's Early White*).—Bunches about five inches long, with a very long stalk, loose, and with many undeveloped berries. Berries medium sized, roundish oval. Skin thin, and green. Flesh very thin and watery, and though without much flavour is agreeable and refreshing. It ripens against a wall in the open air, and is well adapted for this mode of cultivation.

St. Peter's. See *Black St. Peter's.*

SCHIRAS (*Ciras; Scyras; Sirrah; Sirac*).—Bunches long, loose, and shouldered. Berries large, oval. Skin thick, reddish-purple, covered with blue bloom. Flesh rather firm and juicy; juice pale red, sugary, and with a delicious aroma. Ripens in a cool vinery; and is as early as the Royal Muscadine.

This fine, large, oval, black grape is that which is grown almost exclusively in the vineyards of the Hermitage, and furnishes the celebrated Hermitage wine. It is said to have been originally introduced from Schiraz, in Persia, by one of the hermits who formerly resided there.

Schwarzer Riessling. See *Black Cluster.*

SCOTCH WHITE CLUSTER (*Blacksmith's White Cluster; Laan Hâtif; Van der Laan Précoce; Diamant*).—Bunches medium sized, very compact. Berries somewhat oval, or roundish oval. Skin white, covered with thin bloom. Flesh tender and juicy, sweet and richly flavoured. This is a very hardy grape, an excellent bearer, and ripens its fruit against a wall in the open air.

Singleton. See *Catawba.*

Sir A. Pytche's. See *Black Prince.*

Sir W. Rowley's Black. See *Black Frontignan.*

Snow's Muscat Hamburgh. See *Muscat Hamburgh.*

Steward's Black Prince. See *Black Prince.*

Stillward's Sweetwater. See *White Sweetwater.*

Stockwood Park Hamburgh. See *Golden Hamburgh.*

Stoneless Round-berried. See *White Corinth.*

Striped Muscadine. See *Aleppo.*

SYRIAN (*Palestine; Jew's; Terre de la Promise*).—Bunches immensely large, broad-shouldered, and conical. Berries large, oval. Skin thick, greenish-white, changing to pale yellow when quite ripe. Flesh firm and crackling, sweet, and, when well ripened, of good flavour.

This is a very good late grape, and generally produces bunches weighing from 7 lbs. to 10 lbs.; but, to obtain the fruit in its greatest excellence, the vine requires to be grown in a hothouse, and planted in very shallow, dry, sandy soil. Speechly states that he grew a bunch at Welbeck weighing 20 lbs., and measuring 21¾ inches long and 19½ inches across the shoulders. It is a strong grower and an abundant bearer.

Terre de la Promise. See *Syrian.*

Teta de Vaca. See *Cornichon Blanc.*

Tokai Musqué. See *Chasselas Musqué.*

TOKAY.—The Hungarian wine, called Tokay, is not produced from any particular description of grape nor grown in any particular vineyard; the name is applied to all wine grown on the hills of Zemplen, of which Tokay is the chief; and the ground so cultivated extends over seven or eight square leagues of surface. The name Tokay is, therefore, applicable to many varieties of grapes, and it has thus been applied to several varieties in this country. But there is one which, being distinct from all the others, I have described under the name of *White Tokay,* which see.

Tottenham Park Muscat. See *Muscat of Alexandria.*

TREBBIANO (*Trebbiano Bianco; Trebbiano Vero; Erbalus; Ugni Blanc*).—Bunches very large, broad-shouldered, and well set. Berries medium sized, roundish-oval, sometimes oval and sometimes almost round. Skin thick, tough, and membranous, somewhat adhering to the flesh; greenish-white, covered with a very delicate bloom. Flesh firm and crackling, sweet and richly flavoured when well ripened.

This is a late grape, requiring the same heat and treatment as the Muscats, and will hang as late as the end of March. It requires fire heat in September and October to ripen it thoroughly before winter sets in. I have seen bunches of this 14 inches long and 10 inches across.

TRENTHAM BLACK (*Muscat Noir d'Espagne*).—Bunches large, tapering, and shouldered. Berries above medium size, oval. Skin, though not thick, is tough and membranous, separating freely from the flesh, of a jet black colour, and covered with thin bloom. Flesh very melting, abundantly juicy, very rich, sugary, and vinous.

A very excellent grape, ripening with Black Hamburgh; but keeping plump long after the Black Hamburgh shrivels. The vine is a free grower and a good bearer; and Mr. Fleming informs me that it resists powerful sun better than any other variety he knows. It was introduced by Mr. Rivers under the name given as a synonyme; but, not being a Muscat, its present name was adopted.

Turner's Black. See *Esperione.*

Ugni Blanc. See *Trebbiano.*

Ulliade. See *Œillade.*

Uva di tri Volte. See *Ischia.*

Van der Laan Précoce. See *Scotch White Cluster.*

Variegated Chasselas. See *Aleppo.*

VERDELHO.—Bunches rather small, conical, and loose. Berries small, unequal in size, and oval. Skin thin and transparent, yellowish-green, but becoming a fine amber colour when highly ripened, with sometimes markings of russet. Flesh tender, sugary, and richly flavoured.

It is from this grape that the Madeira wine is principally made.

Vert Précoce de Madère. See *Early Green Madeira.*

Victoria Hamburgh. See *Frankenthal.*

Violet Frontignan. See *Purple Constantia.*

Violette Muskateller. See *Purple Constantia.*

Wantage. See *Lombardy.*

Warner's Hamburgh. See *Black Hamburgh.*

Waterzoet Noir. See *Black Sweetwater.*

Weisser Cibebe. See *Early White Malvasia.*

Weisser Muskateller. See *White Frontignan.*

Weisser Riessling. See *White Rissling.*

WEST'S ST. PETER'S (*Black Lombardy, Money's St. Peter's; Poonah; Raisin des Carmes; Raisin de Cuba*).—Bunches large, tapering, and well shouldered. Berries large, roundish-oval, and varying in size. Skin thin, very black, covered with a blue bloom. Flesh tender, very juicy, sweet, and with a fine sprightly flavour.

This is a very fine late grape, and requires to be grown in a house with stove heat.

White Chasselas. See *Royal Muscadine.*

White Constantia. See *White Frontignan.*

WHITE CORINTH (*White Kishmish; Stoneless Round-berried; Corinthe Blanc*).—Bunches small, shouldered, and loose. Berries very small. Skin yellowish-white changing to amber, covered with white bloom. Flesh very juicy, sub-acid and with a refreshing flavour. The seeds are entirely wanting. Of no value.

White Cucumber. See *Cornichon Blanc.*

WHITE FRONTIGNAN (*White Constantia; Nepean's Constantia; Muscat Blanc; Moscatel Commun; Raisin de Frontignan; Weisser Muskateller*).—Bunches large, long, cylindrical, and compact, without shoulders. Berries medium sized, round. Skin dull greenish-white, or yellow, covered with thin grey bloom. Flesh rather firm, juicy, sugary, and very rich, with a fine Muscat flavour.

This will ripen either in a cool or warm vinery, but is worthy of the most favourable situation in which it can be grown. The vine is an abundant bearer, and forces well.

White Hamburgh. See *White Lisbon.*

White Kishmish. See *White Corinth.*

WHITE LISBON (*White Hamburgh; White Portugal; White Raisin*).—Bunches large and loose. Berries oval. Skin greenish-white. Flesh firm and crackling, not very juicy, but with a sweet and refreshing flavour.

It is this grape which is so largely imported from Portugal during the autumn and winter months, and sold in the fruiterers' and grocers' shops under the name of Portugal Grapes.

White Melier. See *Early White Malvasia.*

White Muscadine. See *Royal Muscadine.*

WHITE NICE.—Bunches very large and loose, with several shoulders. Berries medium sized, round, and hanging loosely on the bunches. Skin thin, but tough, and membranous; greenish-white, becoming pale amber coloured as it ripens. Flesh firm and sweet. Bunches of this variety have been grown to weigh 18 lbs. The leaves are very downy underneath.

White Portugal. See *White Lisbon.*

White Raisin. See *White Lisbon.*

WHITE RISSLING (*Weisser Riessling*).—Bunches small, short, and compact, scarcely, if at all, shouldered. Berries round, or somewhat oblate. Skin thin, greenish-white, and, when highly ripened, sometimes with a reddish tinge. Flesh tender, fleshy, and juicy, with a sweet and agreeably aromatic flavour.

This may be grown either in a cool vinery, or against a wall in the open air. The vine is a great bearer, and is very extensively grown in the vineyards of the Rhine and Moselle.

WHITE ROMAIN (*Muscat Romain*).—Bunches below medium size, and rather closely set. Berries medium sized, oval. Skin thin, and so transparent that the seeds can be seen through it; yellowish white, and with a thin bloom. Flesh tender, very juicy, and sweet. An excellent early grape. The wood is very short-jointed, and the vine forms a small bush; it is well suited for pot culture.

Mr. Rivers introduced this variety, expecting it to be a Muscat; but when it fruited it was found not to be so, and he, therefore, adopted the present name.

WHITE SWEETWATER (*Stillward's Sweetwater; Dutch Sweetwater; Perle Blanche*).—Bunches rather above medium size, shouldered, and very loose, containing many badly-developed berries. Berries large and round. Skin thin and transparent, exhibiting the veins of the flesh; white, and covered with a thin bloom, and when highly ripened streaked with traces of russet. Flesh tender, very juicy, sweet, and with a fine delicate flavour.

A well-known and excellent early grape, whose greatest fault is the irregularity with which its bunches are set. There is another Sweetwater, called, by the Dutch, *Water-zoet Witte*, which is a very inferior variety to this.

WHITE TOKAY.—Bunches rather large and compact, from nine inches to a foot long, and broad-shouldered. Berries large and oval. Skin thin, pale coloured, but assuming an amber colour at maturity. Flesh tender and juicy, with a rich flavour. This, in the size of the bunch and form and size of the berries, resembles Muscat of Alexandria; but the bunches are much more compact, and the fruit has not the slightest trace of the Muscat flavour.

Wilmot's Hamburgh. See *Dutch Hamburgh.*

Worksop Manor. See *Black Damascus.*

Zante. See *Black Corinth.*

---

## LIST OF SELECT GRAPES.

*For small establishments those marked * should be chosen.*

### I. FOR WALLS IN THE OPEN AIR.

*Muscats.*

*Early Black Muscat
Early Saumur Muscat
July Muscat
*Muscat St. Laurent

*Not Muscats.*

Black July
*Early Malingre
Early White Malvasia
*Esperione
*Miller's Burgundy
Pitmaston White Cluster
Purple Fontainbleau
*Royal Muscadine

### II. FOR COOL VINERIES.

*Muscats.*

Black Frontignan
Chasselas Musqué
*Early Black Muscat
Early Saumur Muscat
July Muscat
*Madeira Muscat
*Muscat St. Laurent

*Not Muscats.*

*Black Champion
*Black Hamburgh
Black Prince
*Chasselas Vibert
Early White Malvasia
*Golden Hamburgh
Pitmaston White Cluster
*Royal Muscadine
*White Romain

### III. FOR POTS IN ORCHARD-HOUSES.

*Muscats.*

August Muscat
Early Saumur Muscat
*July Muscat
*Muscat St. Laurent
*Sarbelle Muscat

*Not Muscats.*

Chaptal
*Cambridge Botanic Garden
*Chasselas Vibert
*Esperione
Early White Malvasia
Prolific Sweetwater
*Purple Fontainbleau
*Royal Muscadine
*White Romain

## IV. FOR FORCING FOR EARLY CROPS.

*Muscats.*

*Chasselas Musqué
Muscat Hamburgh
Purple Constantia
Red Frontignan
*White Frontignan

*Not Muscats.*

Black Champion
*Black Hamburgh
Black Prince
*Early Chasselas
Golden Hamburgh
*Royal Muscadine
*Trentham Black
White Sweetwater

---

## V. FOR FORCING FOR LATE CROPS.

*Muscats.*

Bowood Muscat
Canon Hall Muscat
*Muscat of Alexandria

*Not Muscats.*

*Barbarossa
Black Damascus
Kempsey Alicante
*Lady Downe's Seedling
*Trebbiano
*West's St. Peter's

---

*The following was accidentally omitted in the alphabetical arrangement.*

Frankenthal (*Black Tripoli; Merrick's Victoria; Pope Hamburgh; Victoria Hamburgh*).—Bunches large and heavily shouldered. Berries roundish, frequently oblate and rarely roundish-oval, sometimes hammered and scarred as in the Dutch Hamburgh. Skin thick, adhering to the flesh, deep black purple, covered with bloom. Flesh firm, and often forming a hollow cell round the seeds, juicy, sugary, sprightly and richly flavoured.

This is very frequently met with in gardens under the name of Black Hamburgh, from which it is easily distinguished by its round, frequently oblate, and hammered berries.

## MEDLARS.

Broad-leaved Dutch. See *Dutch.*

DUTCH (*Broad-leaved Dutch; Gros Fruit; Gros Fruit Monstrueux; Large Dutch*).—This is by far the largest and most generally grown of the cultivated medlars. The fruit is frequently two inches and a half in diameter, and very much flattened. The eye is very open, wide, and unequally rent, extending in some instances even to the margin of outline of the fruit. It is of excellent flavour, but, in that respect, inferior to the following. The young shoots are smooth.

Gros Fruit. See *Dutch.*

Gros Fruit Monstrueux. See *Dutch.*

Large Dutch. See *Dutch.*

Narrow-leaved Dutch. See *Nottingham.*

NOTTINGHAM (*Narrow-leaved Dutch; Small Fruited*).—This is considerably smaller than the Dutch, rarely exceeding an inch and a half in diameter; turbinate, and is more highly flavoured. The young shoots are downy.

Sans Noyau. See *Stoneless.*

Sans Pepins. See *Stoneless.*

Small Fruited. See *Nottingham.*

STONELESS (*Sans Noyau; Sans Pepins*).—In shape this resembles the Nottingham; but it rarely exceeds three quarters of inch in diameter. The eye is smaller and less rent than in the other varieties. It is quite destitute of seeds, and woody core; but the flavour, though good, is inferior to that of the others, being less piquant.

## MULBERRIES.

The only variety cultivated in this country for its fruit is the Black Mulberry (*Morus nigra*), and it is only in the southern counties where it attains perfection. In the midlands it ripens its fruit when trained against a wall; but it is doubtful whether the crop so obtained is sufficient remuneration for the space the tree occupies.

# NECTARINES.

## SYNOPSIS OF NECTARINES.

### I. FREESTONES.

Flesh separating from the stone.

* *Leaves without glands.*

A. *Flowers large.*
Bowden
Hardwicke Seedling

B. *Flowers small.*
Hunt's Tawny

** *Leaves with round glands.*

A. *Flowers large.*
Pitmaston Orange

B. *Flowers small.*
Boston.

*** *Leaves with kidney-shaped glands.*

A. *Flowers large.*
Fairchild's Early
Rivers' Orange
Stanwick
White
Duc du Telliers
Elruge
Impératrice
Late Melting
Murrey
Oldenburg
Peterborough
Violette Grosse
Violette Hâtive

B. *Flowers small.*
Balgowan
Downton

### II. CLINGSTONES.

Flesh adhering closely to the stone.

† *Leaves without glands.*

A. *Flowers large.*
Early Newington
Old Newington

B. *Flowers small.*
None

†† *Leaves with kidney-shaped glands.*

A. *Flowers large.*
Roman

B. *Flowers small.*
Golden

---

Anderdon's. See *Old Newington.*
Aromatic. See *Violette Hâtive.*
BALGOWAN (*Balgone*).—Fruit very large, roundish, in-

clining to ovate. Skin pale green, mottled with red on the shaded side; but entirely covered with deep, bright red on the side next the sun. Flesh with a greenish tinge, veined with red at the stone, melting, very rich, and highly flavoured. Flowers small. Glands kidney-shaped.

A very excellent variety, nearly allied to Violette Hâtive, but is much hardier and a more vigorous grower than that variety. It ripens in the end of August and beginning of September.

Black. See *Early Newington.*

Black Murrey. See *Murrey.*

BOSTON (*Lewis'; Perkins' Seedling*).—Fruit very large, roundish oval. Skin bright yellow on the shaded side, and deep red on the side next the sun. Flesh yellow, without any red at the stone, with an agreeable, but not rich, flavour. Flowers small. Glands round.

Remarkable only for the size and beauty of the fruit, which ripens in the middle of September; and requires a warm season to bring it to maturity. It is an American variety.

BOWDEN.—This is a very large variety, of a round shape. Skin greenish on the shaded side, dark red next the sun, and with a disposition to be russety. The flesh is melting, rich, and sugary, with a slightly astringent flavour. Glands none. Flowers large. Ripe in August. The tree is a very dwarf and compact grower.

Brinion. See *Violette Hâtive.*

Brugnon Musqué. See *Roman.*

Brugnon Red-at-stone. See *Violette Hâtive.*

Claremont. See *Elruge.*

DOWNTON.—Fruit rather larger than Violette Hâtive, roundish oval. Skin pale green in the shade, but deep red next the sun. Flesh pale green, reddish at the stone, melting, juicy, and richly flavoured. Glands kidney-shaped. Flowers small.

A first-rate variety, ripe in the end of August and beginning of September. The tree is a vigorous grower, and an excellent bearer. It was raised by Mr. Knight from the Elruge and Violette Hâtive.

DUC DU TELLIERS (*Duc de Tello; Dutilly's*).—This variety bears a close resemblance to Elruge, with which it is, by some, considered synonymous. It is, no doubt, another form of that variety, and differs only in the greater hardiness and vigour of the tree. Glands kidney-shaped. Flowers small.

Early Black. See *Early Newington.*

Early Newington (*Black; Early Black; Lucombe's Black; Lucombe's Seedling; New Dark Newington*).—Fruit large, roundish ovate, enlarged on one side of the suture; apex ending in a swollen point. Skin pale green in the shade, but bright red, marbled with deeper red next the sun, covered with a thin bloom. Flesh greenish-white, very red next the stone, to which it adheres; rich, sugary, vinous, and very excellent. Earlier and much richer than the Old Newington. Flowers large. Glands none. Ripens early in September.

Early Violet. See *Violette Hâtive.*

Elruge (*Claremont; Oatlands; Springrove; Temple*).—Fruit medium sized, roundish oval. Skin pale greenish in the shade, deep red next the sun, interspersed with dark brownish russet specks. Flesh pale green, reddish towards the stone, melting, juicy, and richly flavoured. Stone oval and rough. Flowers small. Glands kidney-shaped. Ripens in the end of August and beginning of September. This is one of the very best nectarines. The tree is an excellent bearer, and forces well.

Emmerton's White. See *White.*

Fairchild's.—Fruit small, round, slightly flattened at the top. Skin yellowish-green, bright red next the sun. Flesh yellow to the stone, dry, and sweet. Stone nearly smooth. Flowers large. Glands kidney-shaped. Ripens in the beginning and middle of August; but it is of little merit, its only recommendation being its earliness.

Flanders. See *White.*

French Newington. See *Old Newington.*

Golden (*Orange*).—Fruit medium sized, roundish-ovate. Skin fine waxen yellow in the shade, and bright scarlet, streaked with red, where exposed. Flesh yellow, adhering to the stone, juicy, and sweet. Flowers small. Glands kidney-shaped. Early in September.

Grosse Violette Hâtive. See *Violette Grosse.*

Hampton Court. See *Violette Hâtive.*

Hardwicke Seedling.—Fruit very large, almost

round, and sometimes inclining to oval. Skin pale green on the shaded side, entirely covered with dark purplish-red next the sun. Flesh greenish, with a tinge of red next the stone, melting, juicy, rich, and highly flavoured. Glands none. Flowers large. Ripens in the middle and end of August.

This was raised from the Elruge at Hardwicke House, near Bury St. Edmunds, and is one of the hardiest and most prolific of nectarines.

Hunt's Tawny (*Hunt's Early Tawny*).—Fruit rather below medium size, roundish-ovate, narrow towards the top, compressed on the sides, enlarged on one side of the suture. Skin pale orange, deep red next the sun, spotted with russety specks. Flesh deep orange, rich, and juicy. Tree hardy and prolific. Flowers small. Glands none. Ripens in the middle and end of August.

Impératrice.—In size and appearance this has a considerable resemblance to Violette Hâtive; but the flesh is not so red at the stone as in that variety. It is very richly flavoured, and when allowed to hang till it shrivels —a property which few of the Freestone Nectarines possess—it becomes quite a sweetmeat. Glands kidney-shaped. Flowers small. Ripens in the beginning of September. The tree is hardy, and an excellent bearer.

Large Scarlet. See *Violette Hâtive.*

Large White. See *White.*

Late Green. See *Peterborough.*

Late Melting.—This appears to be a variety of Peterborough, but the fruit is double the size. Glands kidney-shaped. Flowers small.

This is a very late variety, and is well worth growing in large collections when it is desired to extend the season of this kind of fruit.

Lewis. See *Boston.*

Lord Selsey's Elruge. See *Violette Hâtive.*

Lucombe's Black. See *Early Newington.*

Lucombe's Seedling. See *Early Newington.*

Murrey (*Black Murrey*). — Fruit medium sized, roundish-ovate, enlarged on one side of the suture. Skin pale green on the shaded side, and dark red next the sun. Flesh greenish-white, melting, and richly flavoured.

Stone nearly smooth. Glands kidney-shaped. Flowers small.

An excellent variety, ripe in the end of August. Tree hardy, and a good bearer.

Neat's White. See *White.*

New Dark Newington. See *Early Newington.*

New Scarlet. See *Violette Hâtive.*

North's Large. See *Old Newington.*

Oatlands. See *Elruge.*

Oldenburg.—Fruit medium sized, ovate. Skin pale yellow on the shaded side, but very much covered with very dark red on the side next the sun. Flesh yellowish-white throughout, and without any trace of red next the stone, very melting and juicy, with a rich, sugary, and vinous flavour. Glands kidney-shaped. Flowers small. Ripens in the end of September, and hangs well till it shrivels, when it is very rich.

Old Newington (*Anderdon's; French Newington; North's Large; Rough Roman; Scarlet Newington; Smith's Newington; Sion Hill*). — Fruit rather large, roundish. Skin pale next the wall, bright red next the sun. Flesh pale yellow, red at the stone, to which it adheres, juicy, sweet, rich, and vinous. Stone small and rough. Flowers large. Glands none. Ripens in the middle of September.

Old Roman. See *Roman.*

Orange. See *Golden.*

Perkins' Seedling. See *Boston.*

Peterborough (*Late Green; Vermash*).—Fruit medium sized, round. Skin green, with a very faint dull red next the sun. Flesh greenish-white to the stone, juicy, but nothing very remarkable except as being the latest nectarine known. Flowers small. Glands kidney-shaped. Ripens in October.

Pitmaston Orange (*Williams' Orange; Williams Seedling*).—Fruit large, roundish-ovate, narrow towards the top, which ends in an acute swollen point. Skin rich orange, brownish-red next the sun, streaked where the two colours blend. Flesh deep yellow, red at the stone, juicy, rich, and excellent. Stone small, sharp-pointed, and very rough. Flowers large. Glands round. Ripens in the end of August and beginning of September. Tree an excellent bearer.

Red Roman. See *Roman.*

RIVERS' ORANGE.—This is a seedling raised from Pitmaston Orange, and differs from its parent in having kidney-shaped instead of round glands. The fruit is similar to that of Pitmaston Orange, and very richly flavoured; and the tree, in Mr. Rivers' estimation, is more robust in its habit, bears, perhaps, more profusely, and is hardier than that variety.

ROMAN (*Brugnon Musqué; Brugnon Violette Musqué; Old Roman; Red Roman*).—Fruit large, roundish, flattened at the top. Skin greenish-yellow, brown muddy red, and rough with russety specks next the sun. Flesh greenish-yellow, deep red at the stone, to which it adheres, rich, juicy, and with a highly vinous flavour, particularly when allowed to hang till it shrivels. Flowers large. Glands kidney-shaped. Beginning of September.

In many collections Violette Hâtive and Elruge are grown for this variety; but from both of these it is readily distinguished by its flowers, which are large.

Rough Roman. See *Old Newington.*

Scarlet. See *Old Newington.*

Sion Hill. See *Old Newington.*

Smith's Newington. See *Old Newington.*

Springrove. See *Elruge.*

STANWICK.—Fruit large, roundish oval. Skin pale lively green where shaded, and purplish-red where exposed to the sun. Flesh white, melting, rich, sugary, and most delicious. Kernel sweet, like that of the sweet almond. Glands kidney-shaped. Flowers large. Ripe the middle and end of September.

The fruit is very apt to crack, and requires to be grown under glass. Hitherto it has generally failed to ripen thoroughly against walls in the open air, except in one or two instances, with which I am acquainted, where grown in a light sandy soil and a good exposure, it then ripened thoroughly without cracking.

Temple. See *Elruge.*

Vermash. See *Peterborough.*

Violet. See *Violette Hâtive.*

Violette de Courson. See *Violette Grosse.*

VIOLETTE GROSSE (*Grosse Violette Hâtive; Violette de Courson*).—Fruit larger than Violette Hâtive. Skin pale green, marbled with violet-red. Flesh less vinous than Violette Hâtive, but an excellent fruit. Flowers small. Glands kidney-shaped. Early in September.

VIOLETTE HÂTIVE (*Aromatic; Early Brugnon; Early Violet; Hampton Court; Large Scarlet; Lord Selsey's Elruge; New Scarlet; Violet; Violette Musqué*).—Fruit large, roundish-ovate. Skin yellowish-green in the shade, dark purplish-red, mottled with brown, next the sun. Flesh yellowish-green, deep red next the stone, rich, sweet, and vinous. Stone roundish, deep reddish-brown, and deeply furrowed. Flowers small. Glands kidney-shaped. Ripens in the end of August and beginning of September.

Violette Musqué. See *Violette Hâtive.*

WHITE (*Emmerton's White; Flanders; Large White; Neat's White; New White; White Cowdray*).— Fruit large, nearly round. Skin white, with a slight tinge of red next the sun. Flesh white throughout, very juicy, with a rich vinous flavour. Stone small. Flowers large. Glands kidney-shaped. Ripens in the end of August and beginning of September.

White Cowdray. See *White.*

Williams' Orange. See *Pitmaston Orange.*

Williams' Seedling. See *Pitmaston Orange.*

---

## LIST OF SELECT NECTARINES.

*Those marked * are suitable for small collections.*

*Balgowan
Downton
Early Newington
*Elruge
*Hardwicke Seedling
Oldenburg
Pitmaston Orange
*Rivers' Orange
*Roman
Stanwick
Violette Hâtive
*White

## NUTS AND FILBERTS.

### SYNOPSIS OF NUTS.

I. NUTS.—*The husk shorter than, or as long as, the Nut.*

Bond
Cob
Cosford
Downton Square
Pearson's Prolific

II. FILBERTS.—*Husk longer than the Nut*

Frizzled
Lambert
Purple
Red
White

---

BOND NUT.—Husk hairy, shorter than the nut. Nut of medium size, ovate and oblong. Shell thin. Kernel large.

This is an excellent nut, and the tree is a good bearer.

Cape Nut. See *Frizzled Filbert.*

COB (*Round Cob*).—Husk hairy, shorter than the nut, and much frizzled. Nut large, obtusely ovate. Shell of a light brown colour, rather thick. Kernel large.

A good nut for early use, but does not keep well.

COSFORD (*Miss Young's; Thin-shelled*).—Husk hairy, as long as the nut, and deeply cut. Nut large, oblong. Shell of a light brown colour, very thin, so much so as to be easily broken between the finger and thumb. Kernel large, and well flavoured.

An excellent early nut, and the tree is an abundant bearer.

DOWNTON SQUARE.—Husk smooth, shorter than the nut. Nut large, short, four-sided. Shell thick. Kernel full, and well flavoured.

Dwarf Prolific. See *Pearson's Prolific.*

Filbert Cob. See *Lambert Filbert.*

FRIZZLED FILBERT (*Frizzled Nut; Cape Nut*).—Husk hairy, twice as long as the nut, deeply frizzled, and spread ing open at the mouth. Nut small, oblong, and flattened. Shell thick. Kernel full.

This is rather a late variety. The tree is an excellent bearer, and the nuts are produced in clusters.

Kentish Cob. See *Lambert Filbert.*

Lambert Filbert (*Kentish Cob; Filbert Cob*).—Husk nearly smooth, longer than the nut, and very slightly cut round the margin. Nut large, oblong, and somewhat compressed. Shell pretty thick, of a brown colour. Kernel full, and very richly flavoured.

This is, perhaps, the best of all the nuts. The tree is a most abundant bearer; some of the nuts are upwards of an inch in length, and they have, with care, been kept for four years. It is only after being kept for some time that their full richness of flavour is obtained.

Miss Young's. See *Cosford.*

Nottingham Prolific. See *Pearson's Prolific.*

Pearson's Prolific (*Dwarf Prolific; Nottingham Prolific*).—Husk hairy, shorter than the nut. Nut medium sized, and smaller than the Cob; obtusely ovate. Shell rather thick. Kernel full.

A very excellent variety. The trees are most abundant bearers, and I have seen them not more than two feet and a half high laden with fruit.

Purple Filbert (*Purple-leaved*).—This differs from the Red Filbert in having the leaves of a dark blood-red colour, like those of the Purple Beech. The fruit is similar to, and quite as good as, that of the Red Filbert, and is of a deep purple colour. It is, therefore, not only valuable as an ornamental shrub, but produces excellent fruit.

Red Filbert (*Red Hazel*).—Husk hairy, longer than the nut. Nut of medium size, ovate. Shell thick. Kernel full, covered with a red skin.

Round Cob. See *Cob.*

Thin-shelled. See *Cosford.*

White Filbert (*Wrotham Park*).—Husk hairy, longer than the nut, round the apex of which it is contracted. Nut medium sized, ovate. Shell thick. Kernel full, and covered with a white skin.

Wrotham Park. See *White Filbert.*

## PEACHES.

### SYNOPSIS OF PEACHES.

#### I. FREESTONES.

Flesh separating freely from the stone.

** Leaves without glands.*

A. *Flowers large.*

Early Anne
Early Savoy
Early York
Hemskerk
Malta
Montaubon
Noblesse
Princesse Marie
Pucelle de Malines
Red Magdalene
Sulhamstead
Vanguard
White Magdalene
White Nutmeg

B. *Flowers small.*

Early Tillotson
Royal Charlotte
Royal George

*** Leaves with round glands.*

A. *Flowers large.*

Abec
Acton Scot
Barrington
Belle Beauce
Early Admirable
Early Grosse Mignonne
Grosse Mignonne
Hâtive de Ferrieres
Leopold the First
Mountaineer
Springrove

B. *Flowers small.*

American Newington
Belle de Doué
Bellegarde
Boudin
Cooledge's Favourite
Crawford's Early
Desse Tardive
George the Fourth
Gregory's Late
Incomparable en Beauté
Late Admirable
Morrisania
Nivette
Têton de Venus
Violette Hâtive
Walburton Admirable
Yellow Alberge

**** Leaves with kidney-shaped glands.*

A. *Flowers large.*

Early Purple
Flat Peach of China
Prince Eugène
Red Nutmeg
Shanghai
Yellow Admirable

### *** *Leaves with kidney-shaped glands.*

B. *Flowers small.*

Belle Chevreuse
Belle de la Croix
Chancellor
Reine des Vergers
Rosanna
Salway
Small Mignonne

## II. CLINGSTONES.

Flesh adhering closely to the stone.

### † *Leaves without glands.*

A. *Flowers large.*

Early Newington
Old Newington

### †† *Leaves with kidney-shaped glands.*

A. *Flowers large.*

Pavie de Pompone

B. *Flowers small.*

Catherine
Incomparable

---

ABEC.—Fruit of medium size and roundish, pitted at the apex, one side of which is higher than the other, and with a shallow suture, which is also higher on one side. Skin remarkably thin and tender, of a lemon-yellow colour, with crimson dots on the shaded side, but covered with a crimson cheek and darker dots of the same colour on the side exposed to the sun. Flesh white, with a very slight tinge of red next the stone, from which it separates very freely; remarkably tender and melting, sweet, and with somewhat of a strawberry flavour. Glands round. Flowers large.

This is a very fine and early peach. It ripens in the third week of August.

Abricotée. See *Yellow Admirable.*

ACTON SCOT.—Fruit small, narrow, and depressed at the top. Skin pale yellowish-white, marbled with bright red next the sun. Suture well marked. Flesh pale throughout, melting, rich, and sugary. Flowers large. Glands round. End of August. A delicious little peach.

Admirable. See *Early Admirable.*

Alberge Jaune. See *Yellow Alberge.*

AMERICAN NEWINGTON (*Early Newington Freestone*). Fruit large, round, and marked with a suture, which is higher on one side. Skin yellowish-white, dotted with red in the shade, and bright red next the sun. Flesh white, red at the stone, to which some strings adhere; juicy, rich, and vinous. Glands round. Flowers small. Ripe in the end of August.

Anne. See *Early Anne.*

Avant. See *Grosse Mignonne.*

Avant Blanche. See *White Nutmeg.*

Avant Pêche de Troyes. See *Red Nutmeg.*

Avant Rouge. See *Red Nutmeg.*

Barrington (*Buckingham Mignonne; Colonel Ansley's*).—Fruit large, roundish-ovate. Skin downy, yellowish-green, marbled with red next the sun. Suture well defined. Flesh yellowish, slightly tinged with red at the stone, rich, vinous, and of first-rate quality. Flowers large. Glands round. Middle of September.

The tree is very hardy, vigorous, and a good bearer. This is one of the best mid-season peaches, and bears carriage well.

Belle Beauce.—This is a variety of Grosse Mignonne, but considerably larger, and ripens from ten to fourteen days later. Glands round. Flowers large. Ripens in the middle of September.

Belle Chevreuse (*Early Chevreuse*).—Fruit elongated, with rarely a nipple on the summit. Skin unusually downy, yellowish, except next the sun, where it is flesh-coloured and marbled with dark red. Suture distinct. Flesh whitish-yellow, tinged with red under the skin next the sun, and marbled with rose colour at the stone, sweet, and juicy. Flowers small. Glands kidney-shaped. Beginning of September.

Belle de la Croix.—This is a new peach raised at Bordeaux eight or ten years ago, and introduced to this country by Mr. Rivers. It is large and round. The flesh is very sweet and richly flavoured, equal to the Early York. Glands kidney-shaped. Flowers small. Ripens in the end of August and beginning of September. The tree is hardy and a robust grower.

Belle de Doué (*Belle de Douai*).—This is a fine, large, melting peach, an early variety of Bellegarde. It is of first-rate quality, with a vinous and richly-flavoured flesh which separates freely from the stone. Glands round. Flowers small. Ripens in the last week of August and beginning of September.

Belle de Paris. See *Malta*.

Bellegarde (*Galande; French Galande; Noire de Montreuil; Ronalds' Brentford Mignonne*).—Fruit round, slightly compressed and hollow at the summit, with a small projecting nipple. Skin deep red all over, striated with dark purple, so much so as to be almost black. Suture shallow. Flesh pale yellow, slightly red at the stone, rich, vinous, and juicy; healthy and a prolific bearer. Flowers small. Glands round. Beginning and middle of September.

This is a very excellent peach, and the tree is a good bearer.

Boudin (*La Royale; Narbonne*).—Fruit large, nearly round, sometimes terminated by a very slight nipple. Skin greenish-white, reddish next the sun, covered with very fine down. Suture deep. Flesh whitish-yellow, deep red round the stone, very rich, sugary, and vinous. Stone small and turgid. Is produced from seed. Flowers small. Glands round. Beginning of September.

Brentford Mignonne. See *Bellegarde*.

Brown Nutmeg. See *Red Nutmeg*.

Buckingham Mignonne, See *Barrington*.

De Burai. See *Yellow Admirable*.

Catherine.—Fruit large, roundish, elongated, swollen on one side of the suture, and terminated by a small nipple. Skin yellowish-green, dotted with bright red in the shade, bright red striated with darker red next the sun. Flesh adhering to the stone, firm, yellowish-white, dark red at the stone, juicy, rich, and excellent. Requires heat to bring it to full perfection. Flowers small. Glands kidney-shaped. September and October.

Chancellor (*Edgar's Late Melting; Late Chancellor; Noisette; Steward's Late Galande*).—Fruit large, oval, pale yellow, dark crimson next the sun. Suture well defined. Flesh free, pale yellow, very deep red at the stone, sugary, rich, and vinous. Stone oblong. Flowers small. Glands kidney-shaped. Middle of September.

China Peach. See *Flat Peach of China*.

Colonel Ansley's. See *Barrington*.

Cooledge's Favourite.—Fruit medium sized, roundish, with a well-defined suture, which is most marked towards the apex, and rather higher on one side than the other. Skin white, covered with crimson dots, and with a crimson cheek on the side next the sun. Flesh very tender and melting, separating freely from the stone, juicy and sweet, and with a fine delicate flavour. Glands round. Flowers small. This is a very fine peach. Ripens in the last week in August.

Crawford's Early (*Crawford's Early Malecoton*).—Fruit very large, of a roundish and slightly oblate shape, depressed at the crown, from which issues a rather shallow suture, much higher on one side than the other. Skin thin, of a deep lemon colour, but on the side next the sun it has a reddish-orange blush, strewed with numerous, distinct, dark crimson dots. Flesh yellow, reddish at the stone, from which it separates freely; very tender and melting, remarkably succulent, with a delicious saccharine and vinous juice. Glands round. Flowers small. End of August and beginning of September.

This is a very large and most delicious peach, with a yellow flesh like an apricot, and is deserving of very extensive cultivation.

Desse Tardive.—Fruit large, round, flat at the top and marked with a deep suture at the stalk. Skin of a very pale colour, covered on the shaded side with minute red dots, and a light tinge of red next the sun. Flesh pale greenish-white, with a faint rosy tinge next the stone, melting, very juicy, sweet, richly flavoured, and vinous. Glands round. Flowers small. Ripe in the end of September and beginning of October.

This is one of the very best late peaches. There is a *Desse Hâtive* quite distinct from this, which ripens in the middle of August, having kidney-shaped glands and large flowers.

Dorsetshire. See *Nivette.*

Double Montagne. See *Montaubon.*

Double Swalsh. See *Royal George.*

Dubbele Zwolsche. See *Royal George.*

Early Admirable (*Admirable*).—Fruit large, roundish. Skin fine clear light yellow in the shade, and bright red next the sun. Suture distinct. Flesh white, pale red at the stone, rich, sweet, and sugary. Flowers large. Glands round. Beginning of September.

EARLY ANNE (*Anne*).—Fruit medium sized, round. Skin white, tinged and dotted with red next the sun. Suture shallow. Flesh white to the stone, pleasant, but rather inclined to be mealy, its earliness proving its chief merit. Flowers large. Glands none. Early in August.

Early Chevreuse. See *Belle Chevreuse*.

EARLY GROSSE MIGNONNE (*Mignonne Hâtive*).—Fruit medium sized, roundish, pitted at the apex, with a small nipple on one side of it, and with a shallow suture. The skin has a pale red cheek on the side exposed to the sun, and is thickly dotted all over with bright crimson dots. The flesh is white, with veins of red throughout, separating freely from the stone, sweet, very juicy, and vinous. Glands round. Flowers large.

This is a very fine peach, ripening in the second week in August.

EARLY NEWINGTON (*Smith's Early Newington*).—Fruit medium sized, rather oval. Skin of a pale straw colour on the shaded side, and streaked with purple next the sun. Flesh pale yellow, tinged with light red next the stone to which it adheres; juicy and well flavoured. Flowers large. Glands none. Ripe the end of August and beginning of September.

Early Newington Freestone. See *American Newington*.

EARLY PURPLE (*Pourprée Hâtive; Pourprée Hâtive à Grandes Fleurs; Vineuse*).—Fruit medium sized, roundish, depressed at the apex, divided on one side by a deep suture extending from the base and across the apex. Skin covered with a thick down, pale sulphur yellow, thinly dotted with red on the shaded side, and deep purplish red next the sun. Flesh white, separating from the stone, red under the skin on the side which is exposed to the sun, and very deep red at the stone; of a rich vinous and sugary flavour. Flowers large. Glands kidney-shaped. Ripe the middle and end of August.

Early Purple Avant. See *Grosse Mignonne*.

EARLY SAVOY (*Précoce de Savoie*).—This is a variety of Grosse Mignonne, but more ovate in shape, and paler colour on the side next the sun. It is an excellent variety, and ripens in the end of August. Glands none. Flowers large.

Early Tillotson.—Very like Royal George.

Early York.—Fruit medium sized, roundish, inclining to ovate; marked on one side with a shallow suture. Skin very thin, delicate greenish-white, dotted with red in the shade, but dark red next the sun. Flesh greenish white melting, very juicy, vinous, and richly flavoured. Glands none. Flowers large. Ripe in the beginning and middle of August.

One of the best early peaches.

Early Vineyard. See *Grosse Mignonne.*

Edgar's Late Melting. See *Chancellor.*

English Galande. See *Violette Hâtive.*

Flat Peach of China (*China Peach; Java Peach*).—Fruit small, so much depressed at both ends as to form a deep hollow on each; in the top one is set a broad, rough, and five-angled eye. Skin pale yellowish-green, mottled with red next the sun. Flesh pale yellow, free, red at the stone, sweet, juicy, and noyeau flavoured; forces well in pots. Middle and end of September. Flowers large. Glands kidney-shaped.

I believe this peach does not now exist in this country. all the trees having been killed by the severe frost of 1838.

Forster's Early. See *Grosse Mignonne.*

French Galande. See *Bellegarde.*

French Magdalen. See *Red Magdalen.*

French Mignonne. See *Grosse Mignonne.*

Galande. See *Bellegarde.*

George the Fourth.—Fruit large, round, swollen on one side of the suture. Skin yellowish-white dotted with red, and rich dark red next the sun, mottled with dark red where the two colours blend. Suture deep at the summit. Stalk set in a hollow depression. Flesh pale yellow, rich, vinous, and juicy. Flowers small. Glands round. Early in September.

This is a very large and very excellent peach.

Golden Fleshed. See *Yellow Alberge.*

Golden Mignonne. See *Yellow Alberge.*

GREGORY'S LATE.—Fruit large, ovate, and pointed. Skin pale green on the shaded side, and with a dark red cheek, like Royal George, on the side next the sun. Flesh very melting, vinous, sugary, and highly flavoured. Glands round. Flowers small. Ripe in the end of September, or beginning of October.

This is an excellent late melting peach, somewhat later than Late Admirable.

Griffith's Mignonne. See *Royal George.*

Grimwood's Royal Charlotte. See *Royal Charlotte.*

Grimwood's Royal George. See *Grosse Mignonne.*

Grosse Jaune. See *Yellow Admirable.*

Grosse Malecoton. See *Pavie de Pompone.*

GROSSE MIGNONNE (*Avant; Early Purple Avant; Early Vineyard; Forster's Early; French Mignonne; Grimwood's Royal George; Johnson's Early Purple Avant; Neil's Early Purple; Padley's Early Purple; Ronalds' Galande; Royal Kensington; Royal Sovereign; Smooth-leaved Royal George; Superb Royal*).—Fruit large, roundish, somewhat flattened, and furrowed with a deep suture at the top, which seems to divide it in two lobes. Skin pale greenish-yellow mottled with red, and deep brownish-red next the sun, covered with fine soft down. Flesh pale yellow, red under the skin on the side next the sun and at the stone, rich, and delicate, vinous, and highly flavoured. Stone small, very rough. Flowers large. Glands round. August and September.

Grosse Pêche Jaune Tardive. See *Yellow Admirable.*

Grosse Persèque Rouge. See *Pavie de Pompone.*

Hardy Galande. See *Violette Hâtive.*

HÂTIVE DE FERRIÈRES.—Fruit medium sized, roundish, marked with a shallow suture, which is higher on one side than the other. Skin white, almost entirely covered with bright red. Flesh white, with a slight tinge of red at the stone, melting and juicy, with a rich vinous flavour. Glands round. Flowers large. Ripens in the beginning of September.

HEMSKERK.—Fruit medium sized. Skin yellowish-green, spotted with scarlet, bright red mottled with darker red next the sun. Flesh greenish-yellow throughout, rich and delicious. Stone small, and smoother than any other peach. A good bearer. Flowers large. Glands none. Ripens in the end of August.

Hoffmann's. See *Morrisania.*

INCOMPARABLE.—Very similar to the Catherine, but not so good. Flesh clingstone. Flowers small. Glands kidney-shaped.

INCOMPARABLE EN BEAUTÉ.—Fruit large, round, and depressed at both ends. Skin pale yellowish-green in the shade, but streaked with crimson and covered with deep brownish-red next the sun. Flesh white, dark red at the stone, melting and juicy, vinous, and with a somewhat musky flavour. A very showy fruit, but is not of first-rate quality. Flowers small. Glands round. Middle of September.

Italian. See *Malta.*

Java Peach. See *Flat Peach of China.*

Johnson's Early Purple Avant. See *Grosse Mignonne.*

Judd's Melting. See *Late Admirable.*

Kew Early Purple. See *Royal Charlotte.*

LATE ADMIRABLE (*Judd's Melting; Motteux' Seedling*).—Fruit very large, elongated, terminated with an acute swollen nipple. Skin yellowish-green, pale red and marbled and striped with deep red next the sun. Suture deep. Flesh greenish-white, with red veins at the stone, delicate, juicy, rich, and vinous. Flowers small. Glands round. Middle and end of September.

One of the best late peaches.

Late Chancellor. See *Chancellor.*

LEOPOLD THE FIRST.—Fruit very large, round, pitted at the apex, and marked with a distinct suture on one side. Skin pale yellow, tinged with red, and very slightly, or not at all, washed with red next the sun. Flesh tender, very melting, vinous, and perfumed. Glands round. Flowers large. Ripens in the middle of October.

Lockyer's Mignonne. See *Royal George.*

Lord Fauconberg's. See *Royal Charlotte.*

Lord Montague's. See *Noblesse.*

Lord Nelson's. See *Royal Charlotte.*

Madeleine Blanche. See *White Magdalen.*

Madeleine de Courson. See *Red Magdalen.*

Madeleine Rouge. See *Red Magdalen.*

Madeleine Rouge à Petites Fleurs. See *Royal George.*

MALTA (*Belle de Paris; Italian; Malte de Normandie; Pêche de Malte*).—Fruit large, roundish, flattened at the top. Skin greenish-yellow, blotched with dull purple next the sun. Suture broad and shallow. Flesh greenish, light red next the stone, rich, vinous, juicy, slightly musky, and deliciously flavoured. Bears carriage better than any other peach. Flowers large. Glands none. August and September.

Mellish's Favourite. See *Noblesse.*

Mignonne Hâtive. See *Early Grosse Mignonne.*

Mignonne Petite. See *Small Mignonne.*

Millet's Mignonne. See *Royal George.*

Monstrous Pavie of Pompone. See *Pavie de Pompone.*

Montagne. See *Montaubon.*

Montagne Blanche. See *White Magdalen.*

MONTAUBON (*Double Montagne; Montagne*).—Fruit medium sized, roundish, narrow at the top. Skin pale greenish-yellow, red, marbled with darker red next the sun. Suture distinct. Flesh white to the stone, rich, and juicy. A good bearer. Flowers large. Glands none. End of August.

MORRISANIA (*Hoffmann's; Morrison's Pound*).—Fruit very large, round. Skin dull greenish-white, and brownish-red next the sun. Flesh pale yellow, juicy, sugary, and richly flavoured. Flowers small. Glands round. Middle and end of September.

Motteux' Seedling. See *Late Admirable.*

MOUNTAINEER. — Fruit large, roundish, somewhat pointed at the apex. Skin nearly smooth, pale yellow, dotted with red on the shaded side, but dark red next the sun. Flesh pale yellowish-green, rayed with red at the stone, melting, juicy, and richly flavoured. Glands round. Flowers large. Early in September.

Neil's Early Purple. See *Grosse Mignonne.*

New Royal Charlotte. See *Royal Charlotte.*

NIVETTE (*Dorsetshire; Veloutée Tardive*).—Fruit round,

elongated, depressed at the top. Skin pale green, bright red with deep red spots next the sun, covered with a fine velvety down. Suture shallow. Flesh pale green, deep red at the stone, rich, and sugary. Flowers small. Glands round. Middle of September.

NOBLESSE (*Lord Montague's; Mellish's Favourite*).—Fruit large, roundish-oblong, terminating with a small nipple. Skin pale yellowish-green in the shade, delicate red, marbled and streaked with dull red and purple next the sun. Flesh white, tinged with yellow, slightly veined with red next the stone, juicy, sweet, and very luscious. Tree hardy and healthy. Flowers large. Glands none. End of August and beginning of September.

Noire de Montreuil. See *Bellegarde.*

Noisette. See *Chancellor.*

Newington. See *Old Newington.*

OLD NEWINGTON (*Newington*).—Fruit large, roundish, marked with a shallow suture. Skin pale yellow in the shade; and fine red marked with still darker red on the side next the sun. Flesh yellowish-white, deep red at the stone, to which it adheres; of a juicy, rich, and very vinous flavour. Flowers large. Glands none. Ripe the middle of September.

D'Orange. See *Yellow Admirable.*

Padley's Early Purple. See *Grosse Mignonne.*

Pavie Camu. See *Pavie de Pompone.*

Pavie Monstrueuse. See *Pavie de Pompone.*

PAVIE DE POMPONE (*Gros Malecoton; Gros Persèque Rouge; Monstrous Pavie of Pompone; Pavie Camu; Pavie Monstrueux; Pavie Rouge de Pompone; Pavie Rouge*).—Fruit immensely large and round, terminated by an obtuse nipple, and marked on one side with a shallow suture. Skin pale yellowish-white, slightly tinged with green on the shaded side, and of a beautiful deep red next the sun. Flesh yellowish-white, deep red at the stone, to which it adheres; in warm seasons it is of a vinous, sugary, and musky flavour, but otherwise it is insipid. In this climate it rarely if ever attains perfection. Flowers large. Glands kidney-shaped. Ripe the middle and end of October.

Pavie Rouge. See *Pavie de Pompone.*

Pavie Rouge de Pompone. See *Pavie de Pompone.*

Pêche d'Abricot. See *Yellow Admirable.*

Pêche de Malte. See *Malta.*

Pêche Jaune. See *Yellow Alberge.*

Petite Mignonne. See *Small Mignonne.*

Petite Rosanne. See *Rosanna.*

Pound. See *Morrisania.*

Pourprée Hâtive. See *Early Purple.*

Pourprée Hâtive à Grandes Fleurs. See *Early Purple.*

Précoce de Savoie. See *Early Savoy.*

PRINCE EUGÈNE.—Fruit medium sized, roundish. Skin pale yellowish-white, and when ripe of a pale waxen colour, faintly tinged with red next the sun. Flesh melting, and somewhat deficient in flavour; but Mr. Rivers says, if forced or grown in a very warm soil and situation it is a very excellent variety. End of August. Glands kidney-shaped. Flowers large.

PRINCESSE MARIE.—Fruit medium sized, roundish. Skin yellowish-white, dotted with pale red on the shaded side, and dark red on the side next the sun. Flesh yellowish-white, rayed with red at the stone; melting, juicy, rich, and vinous. Glands none. Flowers large. Ripens in the middle of September.

PUCELLE DE MALINES.—Fruit pretty large, round, and depressed, having a well-marked suture. Skin clear yellow in the shade, but lightly coloured with red next the sun, and marked with brown spots. Flesh yellowish-white, slightly marbled with red round the stone, melting, juicy, sugary, and with a delicious perfume. Glands round. Flowers large. End of August and beginning of September.

Purple Alberge. See *Yellow Alberge.*

Red Alberge. See *Yellow Alberge.*

Red Avant. See *Red Nutmeg.*

RED MAGDALEN (*French Magdalen; Madeleine de Courson; Madeleine Rouge*).—Fruit rather below medium size, round, and flattened at the stalk. Skin pale yellowish-white in the shade, fine bright red next the sun. Suture deep, extending on one side. Flesh white, veined with red at the stone, firm, rich, sugary, and vinous. Flowers large. Glands none. End of August and beginning of September.

RED NUTMEG (*Avant Pêche de Troyes; Avant Rouge; Brown Nutmeg; Red Avant*).—Fruit small, roundish, terminated by a small round nipple. Skin pale yellow, bright red, marbled with dark vermilion next the sun. Suture distinct. Flesh pale yellow, reddish under the skin on the side next the sun and at the stone, sweet and musky. Very early and hardy. Flowers large. Glands kidney-shaped. July and August.

Valuable only for its earliness.

REINE DES VERGERS.—This is a large, handsome peach, somewhat oval in shape, with a melting flesh of good flavour, but is apt to become pasty unless grown in a warm soil and situation. The tree is very hardy, and, according to Mr. Rivers, succeeds admirably in pots. Ripens in the middle of September. Glands kidney-shaped. Flowers small.

Ronalds' Brentford Mignonne. See *Bellegarde.*

Ronalds' Galande. See *Grosse Mignonne.*

ROSANNA (*Petite Rosanne; St. Laurent Jaune*).—Fruit medium sized, roundish. Skin yellow, deep purplish next the sun. Flesh deep yellow at the circumference, and deep red at the stone; firm, rich, sugary, and vinous. Tree bears well as a standard, and is very productive. Flowers small. Glands kidney-shaped. Middle of September.

This is very different from Alberge Jaune, which is sometimes called Rosanna.

Rouge Paysanne. See *Red Magdalen.*

Royale. See *Boudin.*

ROYAL CHARLOTTE (*Grimwood's Royal Charlotte; Kew Early Purple; Lord Fauconberg's; Lord Nelson's; New Royal Charlotte*).—Fruit rather large, roundish-ovate. Skin pale white, deep red next the sun. Suture moderately distinct. Flesh whitish, pale red next the stone, juicy, rich, and vinous. Flowers small. Glands none. Beginning of September.

ROYAL GEORGE (*Double Swalsh; Dubbele Zwolsche; Griffith's Mignonne; Lockyer's Mignonne; Madeleine Rouge à Petites Fleurs; Millet's Mignonne; Superb*).—

Fruit large, round, and depressed. Skin very pale, speckled with red in the shade, marbled with deeper colour next the sun. Suture deep and broad at the top, extending round almost the whole circumference of the fruit. Flesh pale yellowish-white, very red at the stone, very juicy, rich, and high flavoured. Flowers small. Glands none. August and September.

Royal Kensington. See *Grosse Mignonne.*

Royal Sovereign. See *Grosse Mignonne.*

St. Laurent Jaune. See *Rosanna.*

Salway.—Fruit medium sized, round. Skin of a deep rich yellow colour. Flesh deep orange colour, very melting, juicy, and vinous. Glands kidney-shaped. Flowers small. Ripe in the end of October and beginning of November.

This is a very excellent late variety. The skin and flesh are like those of an apricot, and the latter is very juicy and highly flavoured.

Sandalie Hermaphrodite. See *Yellow Admirable.*

Scandalian. See *Yellow Admirable.*

Shanghai.—Fruit very large, roundish. Skin pale yellowish-green on the shaded side, and light red next the sun. Flesh pale yellow, very deep red at the stone, to which some of the strings adhere; melting, juicy, and richly flavoured. Glands kidney-shaped. Flowers large. Ripens in the middle of September. The tree is an excellent bearer, and requires a very warm situation to ripen the fruit properly. It was introduced from China by Mr. Fortune.

Small Mignonne (*Petite Mignonne; Double de Troyes*).—Fruit small, roundish, flattened at the base, marked on one side with a deep suture. Skin yellowish-white in the shade, and bright red next the sun. Flesh white, pale red next the stone, melting, very juicy, rich, and excellent. Stone small and oblong. Glands kidney-shaped. Flowers small. Ripens early in August.

This ripens after the Red Nutmeg, and is one of the best early peaches. The tree is well adapted for pot culture.

Smith's Early Newington. See *Early Newington.*

Smooth-leaved Royal George. See *Grosse Mignonne.*

SPRINGROVE.—Fruit medium sized. Skin pale green in the shade, bright red next the sun. Excellent, very much resembles Acton Scot. Flowers large. Glands round. End of August and beginning of September.

Steward's Late Galande. See *Chancellor.*

SULHAMSTEAD.—Fruit roundish, depressed. Skin pale yellowish-green, with fine red next the sun. Flesh very excellent. This very much resembles the Noblesse. Flowers large. Glands none. End of August and beginning of September.

Superb. See *Royal George.*

Superb Royal. See *Grosse Mignonne.*

TÊTON DE VENUS.—Fruit elongated, larger than the Boudin, but much paler, having but little colour next the sun, and pale yellowish-white in the shade, surmounted by a large turgid nipple. Flesh white, red at the stone, delicate, sugary, and very rich. Flowers small. Glands round. End of September.

This is quite distinct from Late Admirable.

VANGUARD.—This is a variety of Noblesse, and so similar to it that the fruits cannot be distinguished the one from the other. The only apparent distinction is in the habit of the trees, which in Vanguard is much more robust and hardy than in the Noblesse; and the maiden plants rise with a prominent leader, while the Noblesse makes a roundheaded bush. Glands none. Flowers large.

Veloutée Tardive. See *Nivette.*

VIOLETTE HÂTIVE (*English Galande; Hardy Galande*).—This is evidently a variety of Bellegarde or French Galande, but is not so large in the fruit, and of a paler colour, although it also is of dark red colour next the sun. It may readily be distinguished by nurserymen, as it grows freely on the Muscle, while the Bellegarde requires the Pear-Plum or Brompton stock. It is a large and very excellent peach, ripening in the middle of September. Glands round. Flowers small.

WALBURTON ADMIRABLE.—Fruit large and round. Skin pale yellowish-green on the shaded side, and crimson, mottled with a darker colour, next the sun. Flesh yellowish-white, melting, juicy, rich, and highly flavoured.

Glands round. Flowers small. Ripens in the end of September and beginning of October.

This is one of the best late peaches, and the tree is very hardy and a good bearer.

White Avant. See *White Nutmeg*.

White Magdalene (*Madaleine Blanche; Montagne Blanche*).—Fruit medium sized, roundish, flattened at the base, and divided by a deep suture which extends from the base to the apex, and terminates in a very slight nipple, which is sometimes wanting. Skin easily detached from the flesh, yellowish-white in the shade, and delicately marked with red next the sun. Flesh white, with some yellowish veins running through it, which are tinged with red next the stone, from which it separates; juicy, melting, rich, sugary, and slightly vinous. Flowers large. Glands none. Ripe the middle of August.

White Nutmeg (*Avant Blanche; White Avant*).—Fruit small, roundish, terminated by a pointed nipple, and divided by a deep suture, which extends from the base to the apex. Skin white in the shade and lightly tinged with pale red next the sun. Flesh white even to the stone from which it separates; rich, sugary, and perfumed. Flowers large. Glands none. Ripe the middle of July.

Yellow Admirable (*Abricotée; Admirable Jaune de Burai; Grosse Jaune; Grosse Pêche Jaune Tardive; d'Orange; Pêche d'Abricot; Sandalie Hermaphrodite; Scandalian*).—Fruit very large, roundish, narrowing towards the crown, where it is somewhat flattened, and from which issues a shallow suture, which diminishes towards the base. Skin fine yellow in the shade, and washed with light red on the side next the sun. Flesh firm, deep yellow, tinged with red under the skin, and at the stone, from which it separates; and of a rich sugary flavour resembling both in colour and taste that of an apricot. Flowers large. Glands kidney-shaped. Ripe in the middle and end of October.

Yellow Alberge (*Alberge Jaune; Gold Fleshed; Golden Mignonne; Pêche Jaune; Purple Alberge*).—Fruit medium sized, round, divided by a deep suture which extends from the base to the apex, where it terminates in a considerable depression. Skin adhering to the flesh, covered with fine down, of a deep rich golden yellow on a portion of the shaded side, and deep red on the other, which extends almost over the whole surface of the fruit. Flesh deep yellow, but rich vermilion at the stone, from which it separates, and of a rich vinous flavour. Flowers small. Glands globose. Ripe the beginning of September.

This in favourable situations succeeds well as a standard, and is frequently grown in nurseries under the name of Rosanna, but erroneously.

---

## LIST OF SELECT PEACHES.

***Arranged in the order of ripening.***

Small Mignonne
Early Grosse Mignonne
Early York
Abec
Crawford's Early
Grosse Mignonne
Royal George
Noblesse
Bellegarde
Barrington
Walburton Admirable
Gregory's Late
Desse Tardive
Salway

## PEARS.

Abbé Mongein. See *Uvedale's St. Germain.*

Abondance. See *Amour.*

Achan (*Black Achan; Black Bess; Red Achan; Winter Beurré*).—Fruit medium sized, obovate, flattened towards the eye. Skin varying from pale greenish-yellow, to dark greyish-green, and covered on one side with dull brownish-red. Eye open, set in a slightly depressed basin. Stalk about an inch long. Flesh tender, rich, melting, sugary, and highly perfumed. Ripe in November.

Though an excellent Scotch dessert pear, this is perfectly worthless in the south of England.

Adam's Flesh. See *Chair à Dames.*

Adèle de St. Denis. See *Baronne de Mello.*

Ah! Mon Dieu. See *Amour.*

Albert. See *Beurré d'Amanlis.*

Albertine. — Fruit above medium size, Doyenné shaped. Skin smooth and shining, of a pale lemon colour, strewed with very large russet dots, and with a faint blush of red next the sun. Eye half open, set in a shallow depression. Stalk short and stout. Flesh very tender, melting, and buttery, piquant and perfumed. A first-rate pear, with a slight musky flavour. Ripe in September and October.

Alexandre Bivort.—Fruit rather below medium size, obovate. Skin shining, clear yellow, and covered with pale brown and green dots. Stalk woody, half an inch to three-quarters long. Flesh white, with a reddish tinge, buttery, melting, and very juicy, richly flavoured, and with a high aroma. A first-rate pear. Ripe in the end of December and continues till February. The tree has a bushy habit of growth.

Alexandre Lambré.—Fruit medium sized, round or roundish-oval, uneven in its outline. Skin smooth, greenish-yellow, with sometimes a tinge of red next the sun, and considerably covered with lines and dots of russet. Stalk an inch long, and thick. Flesh white, half-melting, very juicy, sweet, and aromatic. December till February.

Alexandrina.—A medium sized, early, melting pear, which succeeds well on the quince, perfectly hardy, and forming a handsome pyramid. Ripe in September.—*Riv. Cat.*

Althorp Crasanne.—Medium size, roundish-obovate, narrowing rather towards the eye than the stalk. Skin pale green, dotted with russet, brownish next the sun. Stalk an inch and a half long, curved, slender, and moderately depressed. Eye with the segments much divided, set in a slightly plaited shallow basin. Flesh white, buttery, juicy, rich, and perfumed. October to November. A first-rate pear in some situations.

Ambrée Gris. See *Ambrette d'Hiver*.

Ambrette. See *Ambrette d'Hiver*.

Ambrette Grise. See *Ambrette d'Hiver*

Ambrette d'Hiver (*Ambrée Gris; Ambrette; Ambrette Grise; Belle Gabrielle; Trompe Valet*).—Fruit medium sized, roundish, almost oval. Skin pale green, or greenish-grey. Eye small and open, with flat and reflexed segments, and set in a shallow basin. Stalk about an inch long, stout, and inserted in a small cavity. Flesh white when grown on the quince, and tinged with green when grown on the pear stock, rich, melting, and juicy, with an agreeable musky perfume, supposed to resemble that of Ambergris, and from which its name is derived. A very good dessert pear; in season from November to January.

Ambrosia (*Early Beurré*). — Fruit medium sized, roundish, depressed, and rather more swollen on one side than the other. Skin greenish-yellow, covered with grey specks. Eye small, closed with short segments, and set in a wide and rather deep basin. Stalk long and slender, rather deeply inserted. Flesh tender, melting, juicy, and highly perfumed. Ripe in September.

Amiral. See *Arbre Courbé*.

Amiré Joannet (*Joannet; Petit St. Jean; St. Jean; Early Sugar; Harvest Pear*).—Fruit small, regularly pyriform. Skin very smooth, at first of a pale greenish-yellow colour, which changes as it ripens to deep waxen yellow, and with a tinge of red next the sun. Eye open, with stout, erect segments, placed even with the surface. Stalk an inch and a half to an inch and three quarters long, stout and fleshy at the insertion. Flesh white, tender, juicy, sugary, and pleasantly flavoured, but soon becomes mealy.

One of the earliest summer pears. Ripe early in July, and requires to be gathered as it is changing to yellow

Amiré Roux. See *Summer Archduke.*

Amoselle. See *Bergamotte d'Hollande.*

Amour (*Abondance; Ah! Mon Dieu; Belle Fertile*).—Fruit small and obovate. Skin pale yellow or citron in the shade, and fine red covered with darker red dots on the side next the sun. Eye small, scarcely at all depressed, surrounded with a few plaits. Stalk an inch long, curved, and inserted in a swollen cavity. Flesh white, tender, and very juicy, with a rich sugary flavour. Ripens in succession from September onwards, but will not keep longer than a fortnight after being ripe.

Ananas (*Ananas d'Eté*).—Fruit large, obtuse pyriform. Skin yellowish-green, almost entirely covered with rough brown russety dots, and with a brownish tinge next the sun. Eye open, with short stiff segments, and set in a shallow basin. Stalk about an inch and a half long, scarcely at all depressed, but generally with a swelling on one side of it. Flesh delicate, melting, buttery, with a pleasantly-perfumed flavour. Ripe in September.

The tree succeeds well as a standard, and is a good bearer.

Angélique de Bordeaux (*Franc-réal Gros; St. Martial*).—Fruit medium sized, obtuse pyriform. Skin smooth, yellowish-green in the shade and pale brownish-red next the sun; strewed with brown dots. Eye small, set in a narrow and rather shallow basin. Stalk thick, an inch and a half long, fleshy at its insertion. Flesh tender, buttery, juicy, and sugary.

An excellent dessert pear from January to April; but to have it in perfection late in the season it requires to be grown against a wall in a deep, rich soil.

Angleterre d'Hiver. See *Bellissime d'Hiver.*

Angoise. See *Winter Bon Chrétien.*

Angora. See *Uvedale's St. Germain.*

Arbre Courbé (*Amiral; Colmar Charnay*).—Fruit above medium size, oval pyriform. Skin pale green, mottled and dotted with pale brown-russet. Eye open, set in a broad, shallow basin. Stalk three quarters to an inch in length, and stout. Flesh greenish-white, half buttery, juicy, and somewhat astringent. October and November. The tree has crooked branches.

Arbre Superbe. See *Fondante d'Automne.*

Archduke d'Eté. See *Summer Archduke.*

Arteloire. See *St. Germain.*

Aston Town.—Below medium size, roundish-turbinate. Skin greenish-white, thickly dotted with russet; rough, like a Crasanne. Stalk an inch and a half long, straight and slender, inserted without any cavity. Eye small, nearly closed, and in a very shallow basin. Flesh yellowish-white, buttery, perfumed, and high flavoured.

A dessert pear of first-rate quality. Ripe in the end of October and beginning of November. The tree is a vigorous grower, attains a very large size, and bears abundantly.

D'Auch. See *Colmar.*

Auguste Benoît. See *Beurré Benoît.*

Austrasie. See *Jaminette.*

Autumn Bergamot (*Bergamot; English Bergamot; York Bergamot*).—Below medium size, roundish, and flattened. Skin yellowish-green, brownish-red next the sun, dotted with grey-russet. Stalk short and thick, set in a wide, round, hollow. Eye small, placed in a shallow basin. Flesh greenish-white, juicy, melting, exceedingly sugary, and richly flavoured.

A fine old dessert pear, ripe in October. The tree is a vigorous grower, hardy, forms a handsome standard, and is a most abundant bearer.

Autumn Colmar (*De Bavay*).—Medium sized, oblong pyriform, irregular, and uneven. Skin pale yellow, spotted with russet. Stalk an inch long, straight, and placed in a small, uneven cavity. Eye small, set in a very shallow basin. Flesh buttery, gritty at the core, rich, sugary, and perfumed. October to November.

The tree is a good bearer, and succeeds well as a standard.

D'Avril.—Fruit large, pyramidal, uneven in its outline, and considerably bossed round the eye. Skin smooth and shining, of a lively dark green colour, with a dark brown tinge next the sun, and patches of ashy-grey russet on the shaded side; the whole surface covered with very large pale-coloured specks. Flesh crisp, juicy, and sweet. March and April.

Badham's. See *Brown Beurré.*

Bancreif. See *Crawford.*

Banneux. See *Jaminette.*

BARONNE DE MELLO (*Adèle de St. Denis; Beurré Van Mons; His*).—Fruit large, of a curved pyramidal shape. Skin almost entirely covered with dark brown russet, which is thin and smooth. Eye small and open, placed in a very slight depression. Stalk half an inch long, slender, and inserted on the surface of the fruit. Flesh greenish-yellow, fine-grained, melting, and buttery; very juicy, rich, sugary, and with a fine aroma.

An autumn dessert pear of first-rate excellence. Ripe in the end of October, and continues three weeks. The tree is very hardy, an excellent bearer, and succeeds well as a standard or pyramid.

Bartlett. See *Williams' Bon Chrêtien.*

De Bavay. See *Autumn Colmar.*

BEADNELL'S SEEDLING.—Fruit medium sized, turbinate. Skin pale yellowish-green, with a blush of red on the side next the sun, and strewed with grey dots. Eye rather open, set in a shallow depression. Stalk about an inch long. Flesh tender and melting, very juicy and sweet. Ripe in October. Tree hardy, and an excellent bearer.

Beauchamps. See *Bergamotte Cadet.*

Beau de la Cour. See *Conseiller de la Cour.*

Beau Present. See *Jargonelle.*

BEAU PRESENT D'ARTOIS (*Present Royal de Naples*).—Fruit large and pyriform. Skin greenish-yellow, covered with patches and dots of brown russet. Eye small and closed, set in a shallow basin. Stalk about an inch long, slightly depressed. Flesh melting, juicy, sweet, and pretty good flavoured. Ripe in September.

Bedminster Gratioli. See *Jersey Gratioli.*

Bein Armudi. See *Bezi de la Motte.*

Bell Pear. See *Catillac.*

Bell Tongue. See *Windsor.*

Belle Alliance. See *Beurré Sterckmans.*

Belle Andrenne. See *Vicar of Winkfield.*

Belle d'Août. See *Belle de Bruxelles.*

Belle Après Noël. See *Fondante de Noël.*

Belle d'Austrasie. See *Jaminette.*

Belle de Berri. See *Vicar of Winkfield.*

BELLE ET BONNE (*Gracieuse*).—Fruit large, roundish. Skin pale yellowish-green, covered with numerous russety and green spots. Eye open, set in a wide shallow basin. Stalk long and slender, fleshy at the base, and inserted in a narrow cavity. Flesh white, rather coarse, tender, buttery, sweet, and pleasantly flavoured. Ripe in September, but not at all a desirable variety to grow.

BELLE DE BRUXELLES (*Belle d'Août; Belle sans Epines; Bergamotte d'Eté Grosse; Bergamotte des Paysans; Fanfarean*). — Fruit large, abrupt pear-shaped. Skin smooth, of a fine clear lemon-yellow colour, with a tinge of red next the sun, and strewed with freckles of russet. Eye small and half open, set in a shallow basin. Stalk an inch long, without a cavity. Flesh white, tender, juicy, sweet, and perfumed.

A good and handsome summer pear, ripe in the end of August. The tree is very hardy, and a great bearer.

BELLE EPINE DU MAS (*Colmar du Lot; Comte de Limoges; Duc de Bordeaux; Epine Dumas; Epine de Rochechouart*).—Fruit medium sized, pyriform. Skin pale lively green, thickly covered with large dots and patches of brown russet on the shaded side; but next the sun marked with reddish-brown and orange. Eye small and open, set in a deep and furrowed basin. Stalk an inch long, stout, and inserted in a deep cavity prominently knobbed round the margin. Flesh tender, half melting, juicy, and sweet, but with little flavour. In use during November and December.

Belle d'Esquerme. See *Jalousie de Fontenay.*

Belle Excellente. See *Duc de Brabant.*

Belle Fertile. See *Amour.*

Belle de Flandres. See *Flemish Beauty.*

Belle Gabrielle. See *Ambrette d'Hiver.*

Belle Heloïse. See *Vicar of Winkfield.*

Belle de Jersey. See *Uvedale's St. Germain.*

BELLE JULIE.—Fruit rather below medium size, oval. Skin clear olive-green, with a faint tinge of dull red on the side next the sun, and considerably marked with russet, particularly round the eye. Eye open, with spreading segments slightly depressed. Stalk an inch long, inserted in a small cavity. Flesh white, buttery, and melting, juicy, sugary, and with a fine aroma.

An excellent pear, ripe during November. The tree forms a beautiful pyramid, and is a good bearer.

Belle Lucrative. See *Fondante d'Automne.*

Belle de Noël. See *Fondante de Noël.*

Belle Noisette. See *Bellissime d'Hiver.*

Belle sans Epines. See *Hampden's Bergamot.*

Belle Vierge. See *Jargonelle.*

Belle de Zees. See *Bonne d'Ezee.*

Bellissime. See *Windsor.*

BELLISSIME D'HIVER (*Angleterre d'Hiver; Belle Noisette; De Bure; Tèton de Venus*).—Fruit very large, turbinate, flattened on the apex. Skin fine green, changing to brownish-yellow on the shaded side, and fine lively red next the sun; covered all over with russety dots. Eye large, set in an open depressed basin. Stalk an inch long, inserted in an irregular cavity. Flesh white, tender, fine, sweet, mellow, and free from grittiness. A stewing pear, in season from November to April.

BELMONT.—Fruit large, obovate. Skin yellowish-green, tinged with brown next the sun, and covered with dots. Stalk very long, slender, and curved. Flesh coarse, but sweet, and juicy. I have found this one of the best stewing or baking pears in use in November and December.

BENVIE.—Fruit small and obovate. Skin yellowish-green, sometimes tinged and streaked with dingy red next the sun, almost entirely covered with thin grey russet and large russet specks. Eye large and open. Stalk long, fleshy at the base, and obliquely inserted. Flesh yellowish, buttery, juicy, and perfumed.

A Scotch dessert pear of great excellence, ripe in August and Seotember. The tree bears immensely, and attains a great size. The fruit is inferior when grown in the south.

Bequêne Musqué.—Fruit large obovate, and irregular in its outline. Skin pale yellow in the shade, and slightly tinged with dull red next the sun, thickly covered with large patches of grey russet. Eye small and open. Stalk an inch long. Flesh white, gritty, crisp, sweet, and musky.

A stewing pear, in use during November, but it is not one of the best for culinary purposes.

Bergamot. See *Autumn Bergamot.*

Bergamotte d'Alençon. See *Bergamotte d'Hollande.*

Bergamotte d'Austrasie. See *Jaminette.*

Bergamotte de Bugi. See *Easter Bergamot.*

Bergamotte Cadette (*De Cadet; Beauchamps; Beurré Beauchamps*). — Fruit medium sized, obovate. Skin greenish-yellow in the shade, and dull brownish-red next the sun, and marked with patches and large dots of pale brown russet. Eye open, set in a wide and rather deep basin. Stalk three quarters of an inch long, inserted in a small cavity. Flesh white, tender, melting, and very juicy, with a rich sugary and musky flavour.

An excellent dessert pear, in use from October to December.

Bergamotte Crasanne. See *Crasanne.*

Bergamotte Dussart.—Fruit rather above medium size, turbinate. Skin lemon-yellow when ripe, strewed with greenish and grey dots over the surface, and a few traces of russet. Eye frequently wanting. Stalk half an inch long. Flesh half-melting, very juicy, sweet, and vinous. December and January.

Bergamotte Esperen.—Fruit medium sized, turbinate, and uneven in its outline. Skin rough from being entirely covered with dark brown russet. Eye very small, with incurved acute segments. Stalk an inch long, woody, and obliquely inserted. Flesh tender, juicy and melting, sweet and richly flavoured.

This is one of our best late pears, ripening from the end of January up till March and April. The tree forms a handsome pyramid, and is an excellent bearer; but in late situations requires a wall.

Bergamotte d'Eté Grosse. See *Belle de Bruxelles.*

Bergamotte Fièvre. See *Fondante d'Automne.*

Bergamotte Fortunée. See *Fortunée.*

Bergamotte de Fougère. Sée *Bergamotte d'Hollande.*

Bergamotte d'Hiver. See *Easter Beurré.*

BERGAMOTTE D'HOLLANDE (*Amoselle; Bergamotte d'Alençon; Bergamotte de Fougère; Beurré d'Alençon; Lord Cheney's; Holland Bergamot; Winter Green*).—Fruit large, roundish. Skin greenish-yellow, covered with brown russet. Stalk an inch and a half long, slender, curved, set in a shallow, one-sided cavity. Eye small, in a wide, deep basin. Flesh white, crisp, very juicy and sprightly. April till June. Requires a wall.

Bergamotte de Pâques. See *Easter Bergamotte.*

Bergamotte de Paysans. See *Belle de Bruxelles.*

Bergamotte de la Pentecôte. See *Easter Beurré.*

BERGAMOTTE DE STRYKER.—Fruit small, roundish, of an even and regular shape. Skin smooth, and somewhat shining, of a greenish-yellow colour, and marked with russet dots. Eye very large and open, set even with the surface. Stalk three quarters of an inch long, quite green, and inserted without depression. Flesh white, half-melting, and very juicy, sweet, and pleasantly flavoured. Ripe in the end of October.

Bergamotte Tardive. See *Easter Beurré.*

Bergamotte Tardive. See *Colmar.*

Bergamotte de Toulouse. See *Easter Beurré.*

Besidery. See *Bezi d'Heri.*

Beurré d'Albret. See *Fondante d'Automne.*

Beurré d'Alençon. See *Bergamotte d'Hollande.*

BEURRÉ D'AMANLIS (*d'Amanlis; d'Albert; Delbert; Hubard; Kaissoise; Thessoise; Plomgastelle; Wilhelmine*).—Fruit large, obovate. Skin yellowish-green on the shaded side, but washed with brownish-red on the side next the sun, and considerably covered with dots and patches of russet. Eye open, set almost even with the surface. Stalk an inch long, obliquely inserted in a shallow knobbed cavity. Flesh white, melting, very buttery and juicy, with a rich sugary and slightly perfumed flavour.

One of our best autumn pears, ripe in September. The

tree is hardy, with a straggling habit of growth, and is an abundant bearer.

Beurré d'Amanlis Panaché.—This is a variety of the preceding, and differs from it merely in having variegated leaves and fruit striped with yellow or orange bands. It ripens at the same time, and is of the same merit.

Beurré Amboise. See *Brown Beurré.*

Beurré Anglaise. See *Easter Beurré.*

Beurré d'Anjou (*Ne Plus Meuris* of the French).—Fruit large, obtuse-pyriform. Skin greenish-yellow, with sometimes a shade of dull red next the sun, marked with patches of russet, and thickly strewed with brown and crimson dots. Eye small and deeply inserted. Stalk short and stout. Flesh white, rather coarse-grained, but melting and juicy, with a brisk and perfumed flavour. Ripe in December and January.

This is quite distinct from the Ne Plus Meuris of Van Mons. It is not unlike Beurré d'Aremberg in appearance.

Beurré d'Apremont. See *Beurré Bosc.*

Beurré d'Aremberg (*Beurré Deschamps; Beurré des Orphelines; Colmar Deschamps; Delices des Orphelines; Deschamps; Duc d'Aremberg; L'Orpheline; Orpheline d'Enghein; Soldat Laboreur*).—Fruit medium sized, obovate. Skin yellowish-green when ripe, and considerably covered with patches, veins, and dots of cinnamon-coloured russet. Eye small, with short segments, which frequently fall off, and set in a deep hollow. Stalk from half an inch to an inch long, obliquely inserted on the surface of the fruit. Flesh white, melting, buttery, and very juicy, with a rich vinous and perfumed flavour.

A first-rate dessert pear, ripe in December and January. The tree forms a handsome pyramid, and is a good bearer, but is apt to canker in cold soils.

Beurré d'Argenson. See *Passe Colmar.*

Beurré Aurore. See *Beurré de Capiaumont.*

Beurré d'Avranches. See *Louise Bonne of Jersey.*

Beurré Bachelier.—Fruit large and obovate, somewhat irregular in its outline. Skin greenish-yellow, strewed with russety dots. Eye small and closed, set in a shallow basin. Stalk short. Flesh buttery and melting, rich, juicy, sugary, and aromatic.

An excellent pear, ripe in December. The tree is

hardy, forms a handsome pyramid, and is a good bearer.

Beurré Beauchamps. See *Bergamotte Cadette*.

Beurré Beaumont. See *Bezi Vaet*.

Beurré Bennert.—Fruit medium sized, turbinate and irregular in its outline. Skin pale yellow, with a red blush on the side next the sun, and covered with a net-work of russet. Eye small. Stalk an inch long. Flesh juicy, sweet, and aromatic. Ripe in January and February.

Beurré Benoît (*Auguste Benoît; Benoît*).—Fruit medium sized, obovate. Skin pale yellow, strewed with patches and dots of pale brown russet. Eye open, placed in a round and shallow basin. Stalk three quarters of an inch long, inserted in a narrow cavity. Flesh white, fine-grained, melting, and very juicy, sugary, and perfumed.

A good pear, ripe in September.

Beurré Berckmans.—Fruit medium sized, turbinate. Skin of a rich lemon-yellow colour, thickly covered all over with russety specks and dots, but round the stalk and over the crown it is completely covered with a coat of cinnamon-coloured russet. Eye open, set in a round furrowed basin. Stalk an inch long, inserted without depression. Flesh white, tender, fine-grained, juicy, sugary, and richly flavoured.

A handsome and very excellent pear, ripe in November and December. The tree makes a handsome pyramid, and is a good bearer.

Beurré Blanc. See *White Doyenné*.

Beurré Blanc de Jersey. See *Bezi de la Motte*.

Beurré du Bois. See *Flemish Beauty*.

Beurré Bosc (*Beurré d'Apremont; Beurré Rose; Calebasse Bosc; Marianne Nouvelle*).—Fruit large, pyriform. Skin almost entirely covered with thin cinnamon-coloured russet, leaving here and there only a small portion of the yellow ground colour visible. Eye open, placed in a shallow basin. Stalk about an inch and a half long, inserted without depression. Flesh white, melting, and buttery, very juicy, rich, and aromatic.

A dessert pear of first-rate quality, ripe in October and November. The tree is a good bearer; but unless grown against a wall, or in a warm situation, the fruit is apt to be crisp or only half-melting.

Beurré de Bourgogne. See *Flemish Beauty*.

Beurré Bretonneau (*Dr. Bretonneau*).—Fruit large, more or less pyriform. Skin rough, with brown russet, which considerably covers the greenish-yellow ground, and sometimes with a brownish-red on the side next the sun. Eye uneven, set in a moderately deep basin. Stalk an inch long, stout. Flesh yellowish-white, and when it ripens tender, juicy, and well flavoured.

A late dessert pear, in use from March till May; but as it rarely ripens except in very warm summers, the flesh is generally crisp, or at best only half-melting.

Beurré Burnicq.—Fruit medium sized, obovate. Skin rough, from a covering of thick russet, and strewed with grey specks. Stalk half an inch long, inserted in a small cavity. Flesh greenish-white, buttery, and melting, with a powerful aroma. Ripe in the end of October.

Beurré Cambron. See *Glout Morceau*.

Beurré de Capiaumont (*Aurore; Beurré Aurore; Capiaumont; Calebasse Vasse*).—Fruit medium sized, obtuse-pyriform. Skin pale yellow in the shade, almost entirely covered with cinnamon-coloured russet, and numerous grey specks, and orange-red next the sun. Eye large and open, level with the surface. Stalk an inch long, fleshy at the base, inserted without depression. Flesh white, delicate, buttery, and melting, juicy, rich, and sugary.

An excellent autumn pear, ripe in October. The tree is hardy, an abundant bearer, and succeeds well either as a standard or a pyramid.

Beurré des Charneuses. See *Fondante des Charneuses*.

Beurré de Chaumontel. See *Chaumontel*.

Beurré Clairgeau.—Fruit large, curved-pyriform. Skin smooth and shining, of a fine lemon-yellow colour, and with a tinge of orange-red on the side next the sun; it is thickly covered all over with large russety dots and patches of thin delicate russet, particularly round the stalk. Eye small and open, level with the surface. Stalk half an inch long, stout, and rather fleshy, with a swollen lip on one side of it. Flesh white, crisp or half-melting, coarse-grained, juicy, sweet, and slightly musky.

A handsome and showy pear, ripe in November. Its appearance is its greatest recommendation.

Beurré Comice de Toulon. See *Vicar of Winkfield.*

Beurré Copretz.—Fruit below medium size, oval, even, and regularly formed. Skin smooth, of an uniform greenish-yellow colour, covered with large patches and dots of russet. Eye small and open, set in a very shallow basin. Stalk very thick and fleshy, inserted without a cavity. Flesh greenish-white, coarse-grained, juicy, and sugary, but with little flavour. November.

Beurré Curtet. See *Comte de Lamy.*

Beurré Davis. See *Flemish Beauty.*

Beurré Davy. See *Flemish Beauty.*

Beurré Defais.—Fruit large, pyramidal. Skin of a pale golden-yellow colour, dotted with large brown russety dots, and with an orange tinge next the sun. Eye very small and open, sometimes wanting, placed in a deep, narrow cavity. Stalk an inch long, inserted in a cavity. Flesh melting, juicy, sugary, and well flavoured. Ripe in December.

Beurré Delfosse (*Delfosse Bourgmestre; Philippe Delfosse*).—Fruit above medium size, obovate. Skin pale yellow, with a blush of pale red on the side next the sun, and covered with patches and dots of thin russet. Eye closed. Stalk three quarters of an inch long, and slender. Flesh buttery, melting, richly flavoured, and highly aromatic. December and January.

Beurré Derouineau.—Fruit medium sized, obovate. Skin green, changing to yellowish as it ripens on the shaded side, and clouded with brownish-red on the side next the sun. Eye open. Stalk half an inch long, thick and woody. Flesh rather gritty, pretty juicy, sweet and aromatic.

A second-rate pear, ripe in November and December.

Beurré Diel (*Beurré Incomparable; Beurré Magnifique; Beurré Royal; De Trois Tours; Dillen; Gros Dillen; Dorothée Royale; Gratioli d'Hiver; Grosse Dorothée; Melon*).—Fruit varying from medium size to very large; obovate. Skin greenish-yellow, covered with numerous large russety dots, and some markings of russet. Eye open, with erect segments set in an uneven basin. Stalk an inch long, stout, and inserted in an open uneven cavity. Flesh yellowish-white, tender, very buttery and melting, rich, sugary, and aromatic.

A first-rate pear, ripe during October and November. The tree is hardy, and an abundant bearer. Succeeds well as a standard.

Beurré Deschamps. See *Beurré d'Aremberg.*

Beurré Doré. See *Brown Beurré.*

Beurré Drapicz. See *Urbaniste.*

Beurré Duhaume.—Fruit medium sized, roundish, and very much flattened. Skin rough to the feel, covered with thin russet, which is thickly strewed with large russet dots. Eye open, set in an uneven basin. Stalk short, thick, and fleshy, obliquely inserted by the side of a fleshy lip. Flesh buttery, melting, and very juicy, with a rich and vinous flavour.

A very excellent pear, in use from December to February. The tree has a diffuse and bushy habit of growth.

Beurré Duval.—Fruit medium sized or large, of a short pyramidal shape. Skin greenish-yellow, covered with large dark-brown russet freckles, and with a flush of red next the sun. Eye large and open, full of stamens, and set in a wide shallow basin. Stalk obliquely inserted on the end of the fruit. Flesh yellowish, melting and juicy, sugary, and with a fine piquancy.

A very fine and distinct-looking pear, in use during November and December. The tree is hardy, and a good bearer as a pyramid.

Beurré d'Elberg. See *Flemish Beauty.*

Beurré Epine. See *Beurré de Rance.*

Beurré de Flandres. See *Beurré de Rance.*

Beurré Foidard. See *Flemish Beauty.*

Beurré Geerards. See *Gilogil.*

Beurré Giffard.—Fruit about medium sized, pyriform or turbinate. Skin greenish-yellow, mottled with red on the side next the sun. Eye closed, set in a shallow basin. Stalk an inch long, slender, and obliquely inserted on the apex of the fruit. Flesh white, melting, and very juicy, with a vinous and highly aromatic flavour.

An early pear of first-rate quality, ripe in the middle of August.

Beurré Goubault.—Fruit medium sized, roundish, and inclining to turbinate. Skin green, even when ripe. Eye large and open, inserted in a shallow basin. Stalk long and slender, inserted in a small cavity. Flesh melting and juicy, sugary, and with a fine perfumed flavour. Ripe in September.

The tree is an excellent bearer, and the fruit should be watched in ripening, as it does not change from green to yellow.

Beurré Gris. See *Brown Beurré.*

Beurré Gris d'Hiver (*Beurré Gris d'Hiver Nouveau; Beürré de Luçon*).—Fruit large, roundish. Skin entirely covered with thin brown russet, and tinged with brownish-red next the sun. Eye small, set in a very shallow basin. Stalk short and thick, inserted in a small cavity. Flesh white, melting and juicy, sugary and slightly perfumed.

A good late pear when grown in a warm situation, but otherwise coarse-grained and gritty. Ripe from January till March. It is best from a wall.

Beurré Hamecker.—Fruit large and round, bossed about the stalk. Skin greenish-yellow, mottled with brown, covered with patches and dots of fine brown russet. Eye small and open. Stalk an inch long. Flesh buttery, melting, and juicy, sugary and perfumed. Ripe in October and November.

Beurré d'Hardenpont. See *Glou Morceau.*

Beurré Hardy.—Fruit large and pyramidal, of a handsome shape and even outline. Skin shining, yellowish-green, thickly covered with large russet dots, and a coat of brown russet round the stalk and the eye. Eye large and open, set in a shallow basin. Stalk an inch long, stout and fleshy, warted at the base, and inserted without depression. Flesh white, melting and very juicy, sweet and perfumed with a rosewater aroma. Ripe in October.

The tree forms a handsome pyramid, and is a good bearer.

Beurré d'Hiver. See *Chaumontel.*

Beurré d'Hiver de Bruxelles. See *Easter Beurré.*

Beurré Incomparable. See *Beurré Diel.*

Beurré Isambert. See *Brown Beurré.*

Beurré Kennes.—Fruit about medium sized, abrupt pear-shaped, truncated at the stalk end. Skin rather rough to the feel, from a coat of brown russet; on the side next the sun, and over a great part of the shaded side, it is of a vermilion red colour. Eye small and open, set in a wide and shallow basin. Stalk three quarters of an inch long, stout, fleshy at the base, and without a cavity. Flesh yellow, coarse-grained, half-melting, juicy, sweet, and aromatic. Ripe in the end of October.

Beurré de Kent. See *Glou Morceau.*

Beurré Langelier.—Fruit medium sized, obtuse-pyriform. Skin pale greenish-yellow, with a crimson blush on the side next the sun, and covered with numerous russet dots. Eye open, set in a shallow and wide basin. Stalk an inch long, inserted in a small cavity. Flesh tender, buttery and melting, with a rich and vinous flavour.

An excellent pear, ripe during December and January. It requires a warm situation.

Beurré Lefèvre (*Beurré de Mortefontaine*).—Fruit large and obovate, sometimes oval. Skin greenish-yellow on the shaded side, and considerably covered with brown russet; but on the side next the sun it is brownish-orange, shining through a russet coating and marked with a few broken streaks of red. Eye very large and open, set in a deep uneven basin. Stalk an inch long, fleshy at the base, and set on the surface of the fruit. Flesh white, rather gritty at the core, melting and very juicy, richly flavoured, and with a peculiar aroma, which is very agreeable.

A delicious pear, ripe in the middle and end of October. The tree is hardy, and an excellent bearer.

Beurré Léon le Clerc.—Fruit medium sized, obovate. Skin smooth, of a lemon-yellow colour, having a tinge of red on one side, and covered with numerous large russet specks. Eye very small and open, set in a narrow and deep basin. Stalk an inch long, inserted in an uneven and rather deep cavity. Flesh white, melting and juicy, sweet and well flavoured, but without any particular aroma. End of October.

Beurré Lombard. See *Glou Morceau.*

Beurré de Luçon. See *Beurré Gris d'Hiver.*

Beurré Lucratif. See *Fondante d'Automne.*

Beurré Magnifique. See *Beurré Diel.*

Beurré de Malines. See *Winter Nelis.*

Beurré de Mérode. See *Doyenné Boussock.*

Beurré Moiré.—Fruit above medium size, obtuse-pyriform. Skin greenish-yellow, considerably covered with pale bright yellow russet and russety dots. Eye small, set in a shallow basin. Stalk an inch long, stout, inserted in a cavity. Flesh buttery and melting, but not richly flavoured, and with a high perfume. Ripe in November.

Beurré de Mortefontaine. See *Beurré Lefèvre.*

Beurré Nantais (*Beurré de Nantes*).—Fruit large and round. Skin covered with a coat of pale brown russet, like the Brown Beurré, through which a little of the greenish-yellow ground colour appears. Eye very small and open, set in a small and narrow basin. Stalk short, stout, and woody, placed on one side of the axis. Flesh rather coarse-grained, gritty at the core, not melting nor very juicy, but with a sweet and peculiar vinous flavour. November and December.

Beurré Napoléon. See *Napoléon.*

Beurré de Noirchain. See *Beurré de Rance.*

Beurré de Noir Chair. See *Beurré de Rance.*

Beurré des Orphelines. See *Beurré d'Aremberg.*

Beurré de Pâques. See *Easter Beurré.*

Beurré de Paris. See *Jargonelle.*

Beurré de Payence. See *Calebasse.*

Beurré de la Pentecôte. See *Easter Beurré.*

Beurré Picquery. See *Urbaniste.*

Beurré Plat. See *Crasanne.*

Beurré Quetelet. See *Comte de Lamy.*

Beurré de Rance (*Bon Chrêtien de Rans; Beurré Epine; Beurré de Flandres; Beurré de Noirchain; Beurré de Noir Chair; Beurré de Rans; Beurré du Rhin; Hardenpont de Printemps*).—Fruit varying from medium size to large; obtuse-pyriform, blunt, and rounded at the stalk. Skin dark green, and covered with numerous dark brown russety spots. Eye small and open, set in a slight depression. Stalk an inch and a half long, generally obliquely inserted in a wide, shallow cavity.

Flesh greenish-white, buttery, melting, and very juicy, with a rich vinous flavour.

A first-rate and delicious late pear, in use from February till May. The tree is perfectly hardy, and a good bearer. In northern and exposed situations it requires a wall.

Beurré du Rhin. See *Beurré de Rance.*

Beurré du Roi. See *Brown Beurré.*

Beurré Rose. See *Beurré Bosc.*

Beurré Roux. See *Brown Beurré.*

Beurré Royal. See *Beurré Diel.*

Beurré St. Amour. See *Flemish Beauty.*

Beurré St. Nicholas. See *Duchesse d' Orleans.*

Beurré Six.—Fruit large, pyriform, bossed on the surface. Skin smooth, pale green, dotted with green and brown dots, and somewhat russeted. Eye closed. Stalk over an inch long, woody. Flesh white, tender, buttery and melting, rich and sugary, and with a high aroma.

A first-rate pear, ripe in November and December.

Beurré Spence. See *Flemish Beauty.*

Beurré Sterckmans (*Belle Alliance; Calebasse Sterckmans; Doyenné Esterkman*).—Fruit large, turbinate. Skin smooth, of a fine bright grass-green colour on the shaded side, and dull red on the side next the sun, marked with traces of russet. Eye open, set in a wide, shallow basin. Stalk three quarters of an inch long, set in a small round cavity. Flesh white, with a greenish tinge, very melting, buttery and juicy, rich, sugary, and vinous, with a fine aroma.

A first-rate dessert pear, ripe during January and February. The tree is an abundant bearer, succeeds admirably on the quince, and forms a handsome pyramid.

Beurré Superfin. — Fruit above medium size, roundish-obovate or turbinate. Skin of a beautiful lemon colour, very much covered with thin cinnamon-coloured russet. Eye small and open. Stalk inserted on the apex of the fruit without depression. Flesh very fine grained, buttery, melting, and very juicy, with a brisk piquant flavour, and fine aroma.

A first-rate dessert pear, ripe in the end of September and beginning of October. The tree is a vigorous grower, hardy and prolific, and succeeds well as a standard or pyramid.

Beurré Thuerlinckx (*Thuerlincks*).—This is a large, coarse pear, of a roundish-obovate shape, five to six inches long and four or five broad. The flesh is somewhat tender and juicy, but without any aroma, and very soon becomes mealy. Not worth growing. Ripe in November and December.

Beurré de Terwerenne. See *Brown Beurré.*

Beurré Van Mons. See *Baronne de Mello.*

Beurré de Wetteren.—Fruit large, roundish, inclining to turbinate, widest in the middle and tapering obtusely towards each end, uneven in its outline. Skin bright green and shining; dull red on the side next the sun, and covered with large russet spots. Eye open, deeply set. Stalk an inch long, stout, and deeply inserted. Flesh yellowish, coarse-grained, and soon becomes mealy. A showy and peculiar-looking pear, but of no value. Ripe in October.

Bezi de Bretagne.—This is very similar in appearance to Passe Colmar, to which race it evidently belongs. The flesh is crisp, breaking, and very coarse-grained, very juicy and sweet, and exactly the flavour of Passe Colmar. It is a very good late pear, at least as good as pears generally are in March and April.

Bezi de Caissoy (*Besi de Quessoi; Nutmeg; Petit Beurré d'Hiver; Rousselet d'Anjou; Small Winter Beurré; Wilding of Caissoy; Winter Poplin*).—Fruit produced in clusters, small, roundish-turbinate. Skin rough, and entirely covered with brown russet. Eye open, set almost even with the surface. Stalk half an inch long. Flesh white, tender, buttery, sweet, and aromatic.

A very nice little winter dessert pear, ripening in succession from November till March. The tree attains a large size, and is a most abundant bearer.

Bezi de Chaumontel. See *Chaumontel.*

Bezi d'Echassery. See *Echassery.*

Bezi d'Esperen.—Fruit about medium size, pyriform, and tapering from the bulge to either end. Skin clear, yellowish-green, mottled and shaded with fawn-coloured russet, and with a tinge of deep red. Stalk about an inch long, slender. Eye open, set in a moderately deep basin. Flesh white, melting, and buttery, very juicy, sugary, and perfumed. An excellent pear, ripe in November, but does not keep long.

Bezi Goubault.—Fruit medium sized, roundish-obovate. Skin lemon-yellow, considerably covered with cinnamon-coloured russet, and strewed with numerous russety dots. Eye large and wide open, with broad segments, and very slightly depressed. Stalk slender and woody, set in a very narrow cavity, with a fleshy lip on one side. Flesh tender, half buttery, rather gritty at the core, and with a powerful rose-water aroma. March and April.

Bezi d'Heri (*Bezi Royal; Besidery*).—Fruit medium sized, roundish. Skin thin, smooth, greenish-yellow, with a tinge of red next the sun. Eye open, and set in a small round basin. Stalk slender, an inch and a quarter long, inserted without depression. Flesh white, fine-grained, crisp, rather dry, and with somewhat of a fennel flavour. In use from October to December.

This is one of the best stewing pears; and the flesh is generally smooth and well-flavoured when cooked.

Bezi de Landry. See *Echassery*.

Bezi de la Motte (*Bein Armudi; Beurré Blanc de Jersey*).—Fruit medium sized, roundish, inclining to turbinate. Skin yellowish-green, thickly covered with brown russet dots. Eye small and open. Stalk an inch long. Flesh white, fine-grained, buttery, melting, with a sweet and perfumed flavour. Ripe in October and November.

Bezi de Quessoy. See *Bezi de Caissoy*.

Bezi Royal. See *Bezi d'Heri*.

Bezi Vaet (*Bezi de St. Waast; Bezi de St. Wat*).—Fruit above medium size, roundish, very uneven on its surface, being bossed and knobbed, the general appearance being that of a shortened Chaumontel. Skin greenish-yellow, very much covered with brown russet; and on the exposed side entirely covered with russet. Eye open, with erect segments placed in a deep and uneven basin. Stalk three quarters of an inch long, stout and somewhat fleshy basin. Flesh yellowish-white, crisp and breaking, very juicy and sweet, with a pleasant aroma, the flavour being very much like that of the Chaumontel.

A first-rate dessert pear, ripe in December and January. Though not richly flavoured, it is so juicy and refreshing as to be like eating sugared ice.

Bishop's Thumb.—Fruit large and oblong. Skin yellowish-green, covered with largo russety dots, and with a rusty red colour on one side. Eye small and open, with long reflexed segments. Stalk one inch long, fleshy at the base, and obliquely inserted. Flesh greenish-yellow, melting and juicy, with a rich sugary and vinous flavour.

An old-fashioned and very excellent dessert pear, ripe in October. The tree is hardy, an abundant bearer, and succeeds well as a standard.

Black Achan. See *Achan.*

Black Bess. See *Achan.*

Black Beurré. See *Verulam.*

Black Worcester (*Parkinson's Warden; Pound Pear*).—Fruit large and obovate. Skin green, entirely covered with rather rough brown russet, and with a dull red tinge next the sun. Eye small and open. Stalk an inch long. Flesh hard, crisp, coarse-grained, and gritty.

A stewing pear, in use from November till February.

Bloodgood.—Fruit medium sized, turbinate, inclining to obovate. Skin yellow, strewed with russety dots and russet network. Eye open, with stout segments. Stalk obliquely inserted. Flesh yellowish-white, buttery and melting, sweet, sugary, and aromatic.

An American pear of good quality, ripe early in August. The tree bears well, and, being so early, is well worth growing.

Bô de la Cour. See *Conseiller de la Cour.*

Bolivar. See *Uvedale's St. Germain.*

Bonaparte. See *Napoléon.*

Bon Chrêtien d'Amiens. See *Catillac.*

Bon Chrêtien d'Espagne. See *Spanish Bon Chrêtien.*

Bon Chrêtien Fondante.—Fruit large, oblong, and regularly formed. Skin green, covered with a considerable quantity of russet, and marked with numerous russety dots on the shaded side, but covered with dark brownish-red next the sun. Eye small and closed. Stalk three quarters of an inch long. Flesh yellowish-white, very melting and very juicy; the juice rather thin, and not highly flavoured, but very cool, pleasant, and refreshing.

A very nice pear, ripe during October and November. The tree bears well as a standard.

Bon Chrêtien d'Hiver. See *Winter Bon Chrêtien.*

Bon Chrêtien Napoléon. See *Napoléon.*

Bon Chrêtien Nouvelle. See *Flemish Bon Chrêtien.*

Bon Chrêtien de Rans. See *Beurré de Rance.*

Bon Chrêtien de Tours. See *Winter Bon Chrêtien.*

Bon Chrêtien Turc. See *Flemish Bon Chrêtien.*

Bon Chrêtien de Vernoise. See *Flemish Bon Chrêtien.*

Bon Papa. See *Vicar of Winkfield.*

Bonne d'Avranches. See *Louise Bonne of Jersey.*

Bonne Ente. See *White Doyenné.*

Bonne d'Ezée (*Belle de Zées; Bonne de Zées*).—Fruit large, pyramidal. Skin straw, with a tinge of green, and thickly marked with traces of brown russet. Eye open, with long linear segments. Stalk slender, an inch long, and obliquely inserted. Flesh white, coarse-grained, and inclining to gritty, half-melting and juicy, with an agreeable perfume.

This is only a second-rate pear, the texture of the flesh being coarse. Ripe in October.

Bonne de Kienzheim. See *Vallée Franche.*

Bonne de Longueval. See *Louise Bonne of Jersey.*

Bonne Louise d'Avranches. See *Louise Bonne of Jersey.*

Bonne Malinaise. See *Winter Nelis.*

Bonne de Malines. See *Winter Nelis.*

Bonne de Noël. See *Fondante de Noël.*

Bonne Rouge. See *Gansel's Bergamot.*

Bonnissime de la Sarthe. See *Figue de Naples.*

De Bordeaux. See *Bezi d'Heri.*

Bosc Sire. See *Flemish Beauty.*

Boss Père. See *Flemish Beauty.*

Bouche Nouvelle. See *Flemish Beauty*.

Braddick's Field Standard. See *Marie Louise*.

Brilliant. See *Flemish Beauty*.

Brocas' Bergamot. See *Gansel's Bergamot*.

BROOMPARK.—Fruit medium sized, roundish-obovate. Skin yellow, sprinkled with cinnamon-coloured russet. Eye small, dry and horny, set in a slight depression. Stalk an inch long. Flesh yellowish, melting, juicy and sugary, with a rich musky flavour.

An excellent dessert pear, ripe in January. The tree is very hardy and vigorous, an excellent bearer, and succeeds well either as a dwarf or standard.

BROUGH BERGAMOT.—Fruit small, roundish-turbinate, tapering into the stalk. Skin rough, being entirely covered with brown russet, except in patches where the green ground colour is visible; on the side next the sun it is tinged with dull red. Eye open, with short-stunted segments. Stalk half an inch long, not depressed. Flesh yellowish-white, rather coarse-grained, but very juicy and sugary, with a rich and highly perfumed flavour.

An excellent pear for the north of England, ripening during December.

BROUGHAM.—Fruit medium sized, roundish-obovate, inclining to oval or ovate. Skin rather rough to the feel, yellowish-green, and covered with large brown russet specks. Eye clove-like, full of stamens, set in a shallow and plated basin. Stalk an inch and a quarter long, and slender. Flesh yellowish-white, tender, and juicy, but somewhat mealy, and having the flavour of the Swan's Egg.

A second-rate pear, ripe in November. The tree is a great bearer.

Brown Admiral. See *Summer Archduke*.

BROWN BEURRÉ (*Beurré Gris; Beurré Dorée; Beurré d'Amboise; Beurré Roux; Beurré du Roi; Beurré de Terwerenne; Badham's; Isambert le Bon*).—Fruit large, obovate. Skin yellowish-green, almost entirely covered with thin brown russet, and faintly tinged with reddish-brown on the side next the sun. Eye small and open, set in an even, shallow basin. Stalk an inch long, set in a small, round cavity, with generally a fleshy lip on one side. Flesh greenish-white under the skin, but yellowish at the centre, tender, buttery, with a rich piquant flavour and musky aroma.

A well-known pear of first-rate excellence, ripe in October. The tree requires to be grown against a wall to have the fruit in perfection; but it succeeds very well as a dwarf in a warm situation.

Buchanan's Spring Beurré. See *Verulam.*

Bujaleuf. See *Virgouleuse.*

Bujiarda. See *Summer Thorn.*

De Bure. See *Bellissime d'Hiver.*

Burgermeester.—Fruit large, oblong or pyramidal, curved, and very uneven on the surface; round at the apex and knobbed about the stalk. Skin yellowish-green, entirely covered with rough russet. Eye very small, set in a shallow basin. Stalk an inch long, obliquely inserted. Flesh yellowish, melting, juicy and sweet, with a fine musky flavour.

A good second-rate pear, ripe in November.

De Cadet. See *Bergamotte Cadet.*

Caillot Rosat (*English Caillot Rosat; King Pear*).—Fruit above medium size, pyriform. Skin smooth, greenish-yellow in the shade, and quite covered with a brownish-red cheek and streaks of brighter red on the side next the sun. Eye open, set in a shallow cavity. Stalk three quarters of an inch long. Flesh tender, very juicy and melting, sweet, and nicely perfumed.

A nice early pear, ripe in August; and the tree is an excellent bearer. This is not the Caillot Rosat of the French, which is the same as our *Summer Rose.*

Calebasse (*Beurré de Payence; Calebasse d'Hollande; Pitt's Calebasse*).—Fruit medium sized, oblong, undulating in its outline. Skin yellow, covered with thin grey russet in the shade, and cinnamon-coloured russet next the sun. Eye open. Stalk an inch and a half long, obliquely inserted, with a fleshy lip on one side. Flesh crisp, juicy, rich, and sugary. Ripe in October.

Calebasse Bosc. See *Beurré Bosc.*

Calebasse Carafon. See *Calebasse Grosse.*

Calebasse Delvigne. — Fruit above medium size, pyriform. Skin yellow, strewed with cinnamon-coloured russet. Eye open, with stout segments, set in a shallow basin. Stalk short and stout. Flesh white, rather coarse-grained, juicy, melting, with a rich and perfumed flavour. Ripe in October.

Calebasse d'Été.—Fruit medium sized, pyramidal. Skin yellow, covered with brown russet. Flesh white, half-melting, very juicy and sweet. A good early pear, ripe in September.

Calebasse Grosse (*Calebasse Carafon; Calebasse Monstre; Calebasse Monstrueuse du Nord; Calebasse Royale; Triomphe de Hasselt; Van Marum*).—Fruit very large, sometimes measuring six inches long, pyramidal. Skin greenish-yellow, considerably covered with dark grey russet in the shade, and entirely covered with light brown russet on the side next the sun. Eye small, set in a pretty deep basin. Stalk an inch long. Flesh coarse-grained, crisp, juicy, and sweet. Ripe in October. Its size is its only recommendation.

Calebasse d'Hollande. See *Calebasse.*

Calebasse Monstre. See *Calebasse Grosse.*

Calebasse Monstrueuse de Nord. See *Calebasse Grosse.*

Calebasse Royale. See *Calebasse Grosse.*

Calebasse Sterckmans. See *Beurré Sterckmans.*

Calebasse Tougard. — Fruit medium sized, sometimes large, pyramidal and curved, uneven in its outline. Skin greenish-yellow, entirely covered with brown russet. Stalk short and thick. Flesh yellowish-white, crisp, juicy, and sweet. Ripe in October and November.

Calebasse Vasse. See *Beurré de Capiaumont.*

De Cambron. See *Glou Morceau.*

Canet. See *Duc de Nemours.*

Canning. See *Easter Beurré.*

Canning d'Hiver. See *Easter Beurré.*

Capiaumont. See *Beurré de Capiaumont.*

Captif de St. Helène. See *Napoléon.*

Cassante de Mars.—Fruit produced in clusters, below medium size, roundish-obovate. Skin deep yellow, speckled and traced with light brown russet. Eye large, and wide open. Stalk about an inch long, inserted without depression. Flesh yellowish-white, crisp and breaking, juicy, sweet, and richly flavoured.

An excellent pear for so late in the season. Ripe in April and May.

CATILLAC (*Bon Chrêtien d'Amiens; Chartreuse; Grand Monarque; Gros Gilot; Monstrueuse de Landes; Têton de Venus; Bell Pear; Pound Pear*).—Fruit very large, flatly turbinate. Skin at first of a pale green colour, changing to lemon-yellow, with a tinge of brownish-red next the sun, and covered with numerous large russet specks. Eye open. Stalk an inch and a half long. Flesh white, crisp, gritty, and with a musky flavour.

One of the best stewing pears, in use from December to April.

CATINKA.—Fruit medium sized, obovate. Skin of a fine deep lemon-yellow colour, thickly covered with large cinnamon-coloured freckles and tracings of russet. Eye rather small, and open. Stalk three quarters of an inch long. Flesh yellowish, melting, but slightly gritty, juicy, very sugary, with a rich full flavour, and a fine aroma of the rose.

A very first-rate pear, with rich saccharine juice, ripe in December.

Cellite. See *Passe Colmar.*

Chambers' Large. See *Uvedale's St. Germain.*

Chambrette. See *Virgouleuse.*

Chapman's. See *Passe Colmar.*

Chapman's Passe Colmar. See *Passe Colmar.*

CHARLES D'AUTRICHE.—Fruit large, roundish, handsome, and regularly formed. Skin greenish-yellow, thickly covered with russety specks and thin patches of grey russet; and with a few streaks of faint red on the side next the sun. Eye open, set in a smooth shallow basin. Stalk an inch long, scarcely at all depressed. Flesh tender, half-buttery and melting, juicy, sugary, and richly flavoured.

A dessert pear, ripe in October. This name is by the French sometimes applied to *Napoléon,* but erroneously.

CHARLES VAN HOOGHTEN.—Fruit large, roundish-oval, even in its outline. Skin of a uniform straw colour, considerably covered with large russety dots, and traces of pale brown russet. Eye wide open. Stalk an inch long, slender. Flesh white, coarse-grained, gritty, half-melting, and not very juicy; sweet, sugary, and rather richly flavoured, and with a musky perfume. Ripe in the end of October and November.

Charlotte de Brouwer.—Fruit large, roundish, inclining to ovate, similar in shape to a large Ne plus Meuris. Skin entirely covered with a coat of light brown russet, with a little of the yellow ground shining through on the shaded side. Eye very small, with short, erect segments. Stalk very short, placed in a knobbed cavity. Flesh white, half-melting, and rather crisp, very juicy, but very astringent. Ripe in October and November.

Charnock (*Drummond; Early Charnock; Scot's Cornuck*).—Fruit small, pyriform. Skin greenish-yellow in the shade, and entirely covered with dark dull red next the sun. Eye small and open. Stalk fleshy, obliquely inserted. Flesh yellowish, half-buttery, juicy, sweet, and with a high aroma.

A Scotch dessert pear, ripe in September, but soon becomes mealy.

Chartreuse. See *Catillac.*

Chaulis. See *Messire Jean.*

Chaumontel (*Bezi de Chaumontel; Beurré de Chaumontel; Beurré d'Hiver; Guernsey Chaumontel; Grey Achan; Oxford Chaumontel; Winter Beurré*).—Fruit large, oblong, or obtuse-pyriform, irregular and undulating in its outline. Skin rather rough, yellowish-green, covered with numerous russety spots and patches, and with brownish-red next the sun. Eye open, set in a deep, irregular basin. Stalk an inch long, inserted in a deep knobbed cavity. Flesh yellowish-white, buttery and melting, rich, sugary, and highly perfumed.

A dessert pear of high merit, in use from November till March.

De Chypre. See *Early Rousselet.*

Citron des Carmes (*Gros St. Jean; Madeleine; Early Rose Angle*).—Fruit below medium size, obovate. Skin smooth and thin, yellowish-green when ripe, and with a faint tinge of brownish-red on the side next the sun. Eye small, and set in a shallow depression. Stalk an inch and a half to two inches long, inserted without depression. Flesh yellowish-white, tender, melting, very juicy and sweet.

A delicious summer pear, ripe in the end of July and beginning of August. It is very apt to crack.

Citron de Septembre. See *White Doyenné.*

COLMAR (*D'Auch; Bergamotte Tardive; Colmar Dorée; De Maune*).—Fruit above medium size, obtuse-pyriform. Skin smooth, pale green, changing to yellowish-green, strewed with grey russet specks. Eye large and open. Stalk an inch to an inch and a half long, stout and curved. Flesh greenish-white, buttery, melting, tender, and with a rich sugary flavour.

An old and justly-esteemed dessert pear, ripening in succession from November till March. The tree is not an abundant bearer, and requires to be grown against a wall.

COLMAR D'AREMBERG (*Fondante de Jaffard; Kartofell*).—Fruit large, obovate, uneven, and bossed in its outline. Skin lemon coloured, marked with spots and patches of russet. Eye rather small and partially closed, set in a very deep round cavity. Stalk short, and rather slender, deeply inserted. Flesh yellowish-white, coarse-grained, half-melting, juicy, and briskly flavoured.

A fine-looking but very coarse pear, ripe in October.

Colmar Charnay. See *Arbre Courbé.*

Colmar Deschamps. See *Beurré d'Arembery.*

Colmar Doré. See *Passe Colmar.*

Colmar Doré. See *Colmar.*

Colmar Epineux. See *Passe Colmar.*

Colmar Hardenpont. See *Passe Colmar.*

Colmar d'Hiver. See *Glou Morceau.*

Colmar Jaminette. See *Jaminette.*

Colmar du Lot. See *Belle Epine du Mas.*

COLMAR NEILL.—Fruit very large, obovate. Skin smooth and glossy, of a uniform yellow colour, dotted and lined with cinnamon-coloured russet. Eye open, set in a wide and rather deep basin. Stalk an inch long, inserted in a small, close cavity. Flesh white, very tender, buttery and juicy, with a high musky flavour. Ripe in October, but soon becomes mealy.

Colmar Nelis. See *Winter Nelis.*

Colmar Preul. See *Passe Colmar.*

Colmar **Souverain.** See *Passe Colmar.*

Colmar Van Mons.—Fruit medium sized, pyramidal, irregular and uneven in its outline. Skin yellowish-green, much covered with a thick coat of smooth brown russet. Eye small and open, set in a small round basin. Stalk three quarters of an inch long, obliquely inserted in a narrow cavity. Flesh yellowish, buttery and melting, very juicy and sweet, but not highly flavoured. Ripe from November to January.

Comte d'Allos.—Large and pyriform, very much the shape of Marie Louise. Skin pale yellow, with a greenish tinge, covered all over with large russety freckles, and with a coating of russet round the eye. Eye very small and open. Stalk three quarters of an inch long. Flesh yellowish, coarse-grained, and rather gritty, melting, juicy, sweet, and richly flavoured, but soon rots at the core. Ripe in December.

Comte de Flandre.—Fruit very large, pyriform. Skin almost entirely covered with large freckles of cinnamon-coloured russet. Eye open, and rather large, with very short deciduous segments. Stalk three quarters of an inch long, slender. Flesh yellowish, melting, juicy, and sugary, with a rich and agreeably perfumed juice.

A first-rate pear, well worth growing, ripe in November and December.

Comte de Lamy (*Beurré Quetelet; Beurré Curtet; Dingler; Marie Louise Nova*).—Fruit medium sized, roundish-obovate. Skin yellowish-green, with brownish-red next the sun, and strewed with russety d ts. Eye small, set in a slight depression. Stalk an inch long, set in a small cavity. Flesh white, tender, buttery, melting, sugary, and richly flavoured.

A delicious pear, ripe in October. Tree hardy, a good bearer, and succeeds well either as a standard or pyramid.

Comte de Limoges. See *Belle Epine du Mas.*

Comtesse de Frênol. See *Figue de Naples.*

Comtesse de Treweren. See *Uvedale's St. Germain.*

Conseiller de la Cour (*Bô de la Cour; Beau de la Cour; Marechal de la Cour*).—Fruit below medium size, pyriform. Skin smooth, yellowish-green, covered with dark green dots, and with a patch of russet round the stalk. Eye large and open, set in a deep, wide hollow. Stalk above an inch long, slender, obliquely inserted, without depression, by the side of a fleshy lip. Flesh white, half-melting, juicy, and briskly flavoured, but not particularly rich. Ripe in January.

Coule Soif. See *Summer Franc Real.*

CRASANNE (*Bergamotte Crasanne; Beurré Plat; Crasanne d'Automne*).—Fruit large, roundish, and flattened. Skin greenish-yellow, marked all over with veins and dots of grey russet. Eye small and open, set in a deep, round, and narrow basin. Stalk two to two inches and a half long, slender and curved, inserted in a small cavity. Flesh white, buttery, melting, rich and sugary, with a fine perfume.

A fine old pear, ripe during November and December. The tree is not a good bearer, and requires to be grown against a wall.

Crasanne d'Austrasie. See *Jaminette.*

Crasanne d'Automne. See *Crasanne.*

Crasanne d'Eté. See *Summer Crasanne.*

CRAWFORD (*Bancrief; Lammas* [of the Scotch]).—Fruit below medium size, obovate. Skin greenish-yellow, changing to pale ye low, with sometimes a tinge of brownish-red next the sun. Eye open. Stalk an inch long. Flesh white, buttery, juicy, sweet, and with a husky flavour. Ripe in the middle of August.

CROFT CASTLE.—Fruit medium sized, oval. Skin greenish-yellow, covered with large brown dots. Eye large and open, with long recurved segments. Stalk an inch and a half long, slender and curved. Flesh very juicy, sweet, and perfumed. Ripe in October.

The tree is a most abundant and regular bearer, succeeds well as a standard, and is well adapted for orchard culture.

Cuiellette. See *Jargonelle.*

Curé. See *Vicar of Winkfield.*

Cypress. See *Early Rousselet.*

Davy. See *Flemish Beauty.*

Dean's. See *White Doyenné.*

DEARBORN'S SEEDLING.—Fruit small, turbinate. Skin

smooth, of a pale yellow colour, strewed with small russety dots. Eye large and open, set in a shallow depression. Stalk long and slender, inserted in a small cavity. Flesh white, very juicy and melting, sweet and pleasantly flavoured. An early pear, ripe in August.

Delbert. See *Beurré d'Amanlis.*

Delfosse Bourgmestre. See *Beurré Delfosse.*

Delices d'Hardenpont.—Fruit above medium size, obtuse-pyriform, irregular and uneven in its outline. Skin smooth, bright lemon-yellow when ripe, thickly covered with pale brown russet. Eye small and open, set in an uneven and considerable depression. Stalk an inch long, thick and fleshy. Flesh white, tender, buttery, and melting, rich, sugary, and perfumed. A good pear, ripe in November.

Delices d'Hardenpont d'Angers.—Fruit medium sized, roundish-obovate. Skin pale yellow, with a tinge of clear red next the sun, strewed with russety dots and patches of russet. Eye small and open. Stalk short and thick, obliquely inserted in a small cavity, and fleshy at the base. Flesh white, rather coarse-grained, juicy, ugary, and agreeably perfumed. Ripe in November.

Delices de Jodoigne.—Fruit medium sized, pyriform. Skin thin, pale yellow, marked with flakes and dots of pale brown russet. Eye open. Stalk short, very thick and fleshy. Flesh half-melting, sweet, sugary, and aromatic. Ripe in the beginning and middle of October.

Deschamps. See *Beurré d'Aremberg.*

Desirée Van Mons. See *Fondante de Charneux.*

Deux Sœurs.—Fruit large, oblong, and ribbed. Skin green, changing to yellowish-green, and strewed with dark dots. Stalk an inch long, curved. Flesh greenish-yellow, buttery, melting, very juicy and sugary. Ripe in November.

Diamant. See *Gansel's Bergamot.*

Dingler. See *Comte de Lamy.*

Dix.—Fruit very large, Calebasse shaped. Skin deep yellow, covered all over with rough russet dots and markings of russet. Eye small, set in a wide, shallow depression. Stalk upwards of an inch in length, stout, and inserted without depression. Flesh rather coarse-grained, juicy, sweet, and slightly perfumed. A second-rate pear, ripe in November.

Dr. Bretonneau. See *Beurré Bretonneau.*

DR. TROUSSEAU. Fruit large and pyriform, wide towards the apex. Skin rough, greenish-yellow, covered with numerous grey specks and russet flakes. Eye open, sometimes wanting. Stalk three quarters of an inch long, woody, and inserted in a narrow cavity. Flesh white, buttery, melting, and very juicy, sugary, and with a powerful aroma. A very excellent pear, ripe in December.

Dr. Udale's Warden. See *Uvedale's St. Germain.*

Dorothée Royale. See *Beurré Diel.*

Double Philippe. See *Doyenné Boussoch.*

Downham Seedling. See *Hacon's Incomparable.*

DOYEN DILLEN.—Fruit above medium size, pyramidal or pyriform. Skin yellow, very much covered with dots and patches of russet. Eye small, half open, and set in a slight depression. Stalk short, thick, and fleshy, inserted without depression. Flesh buttery and melting, very juicy, sweet, and richly flavoured. An excellent pear, ripe in November.

Doyenné d'Automne. See *Red Doyenné.*

Doyenné Blanc. See *White Doyenné.*

DOYENNÉ BOUSSOCH (*Beurré de Mérode; Double Philippe; Nouvelle Boussoch*). — Fruit very large, roundish-obovate, or Doyenné shaped. Skin lemon coloured, covered with large, rough, russety dots. Eye open, placed in a shallow basin. Stalk short and stout, inserted in a narrow cavity. Flesh yellowish-white, tender, very melting and juicy, with a fine brisk vinous juice, and a delicate, agreeable perfume.

A delicious and very handsome pear, ripe in October.

Doyenneé Crotté. See *Red Doyenné.*

DOYENNÉ DEFAIS.—Fruit small, roundish-obovate, or Doyenné-shaped, bossed at the stalk end. Skin yellow, very much covered with cinnamon-coloured russet. Eye rather large and wide open, set in a shallow depression. Stalk about an inch long, set in a deep, wide, and furrowed cavity. Flesh tender, buttery, melting, and very juicy, rich, sugary, and vinous, with a fine musky aroma.

A most delicious pear; one of the best. Ripe in December. The tree is hardy, and a good bearer.

Doyenné Esterckman. See *Beurré Sterckmans.*

Doyenné d'Eté. See *Summer Doyenné.*

Doyenné Galloux. See *Red Doyenné.*

Doyenné Gris. See *Red Doyenné.*

DOYENNÉ GOUBAULT.—Fruit above medium size, obovate, inclining to pyriform. Skin pale yellow, with markings of russet about the stalk and the eye, and covered with russety dots. Eye small, set in a rather deep hollow. Stalk short and thick. Flesh melting, juicy, rich, sugary, and aromatic. An excellent pear, ripe in January.

Doyenné d'Hiver. See *Easter Beurré.*

Doyenné d'Hiver Nouveau. See *Easter Beurré.*

Doyenné Jaune. See *Red Doyenné.*

Doyenné de Juillet. See *Summer Doyenné.*

Doyenné de Pâques. See *Easter Beurré.*

Doyenné Pictée. See *White Doyenné.*

Doyenné de Printemps. See *Easter Beurré.*

Doyenné Rouge. See *Red Doyenné.*

Doyenné Roux. See *Red Doyenné.*

Drummond. See *Charnock.*

Dry Martin. See *Martin sec.*

Duc d'Aremberg. See *Beurré d'Aremberg.*

Duc de Bordeaux. See *Belle Epine du Mas.*

Duc de Brabant. See *Fondante de Charneux.*

DUC DE NEMOURS (*Canet*).—Fruit growing in clusters, oblong-obovate. Skin yellow, strewed with reddish and grey dots. Eye open. Stalk an inch long, thick, inserted without depression on a fleshy knob. Flesh rather coarse-grained, juicy and sweet. A second-rate pear, ripe in December.

Duchesse. See *Duchesse d'Angoulême.*

DUCHESSE D'ANGOULÊME (*Duchesse; Eparonnais; de*

*Pézénas*).—Fruit large, sometimes very large, roundish-obovate, very uneven and bossed on its outline. Skin pale dull yellow, covered with veins and freckles of pale brown russet. Eye open, set in a deep, irregular basin. Stalk an inch and a half long, stout, inserted in a deep irregular cavity. Flesh white, buttery, and melting, with a rich flavour when well ripened; but generally it is coarse-grained and half-melting, juicy, and sweet.

A dessert pear, sometimes of great excellence, ripe during October and November.

Duchesse de Berri d'Eté. See *Summer Doyenné.*

Duchesse de Berri d'Hiver. See *Uvedale's St. Germain.*

DUCHESSE DE BRABANT.—Fruit medium sized, short pyriform, even in its outline. Skin very thin, smooth and shining, greenish-yellow, thickly strewed with russety dots, and with a patch of russet round the eye. Eye large and open, set in a shallow basin. Stalk an inch long, inserted without depression. Flesh yellowish-white, buttery and melting, very juicy and sweet, with a pleasant aroma.

An agreeable and refreshing pear, ripe in November.

DUCHESSE DE MARS.—Fruit medium sized, obovate. Skin yellow, with a tinge of reddish-brown next the sun, and considerably covered with brown russet. Eye small and closed, set in a shallow depression. Stalk an inch long, inserted without depression. Flesh buttery, melting, juicy, and well flavoured. Ripe in November.

DUCHESSE D'ORLEANS (*Beurré St. Nicholas; St. Nicholas*).—Fruit large and pyriform. Skin yellow on the shaded side, but with a tinge of red on the side next the sun, mottled with greenish-brown russet. Eye open, set in a wide, shallow basin. Stalk three quarters of an inch long. Flesh yellowish-white, melting, buttery and juicy, with a rich, sugary, and vinous flavour, and fine aroma. A most delicious pear, ripe in October.

DUNMORE.—Fruit large, oblong-obovate. Skin greenish, marked with numerous dots and patches of brown russet, and with a brownish-red tinge next the sun. Eye small and open, set in a rather deep and narrow basin. Stalk an inch and a half long. Flesh yellowish-white, buttery, and melting, with a rich sugary flavour. Ripe in September and October.

Early Beurré. See *Ambrosia.*

Early Catherine (of America). See *Early Rousselet.*

Early Charnock. See *Charnock.*

Early Rose Angle. See *Citron des Carmes.*

Early Rousselet (*De Chypre; Cypress; Early Catherine; Perdreaux; Perdreaux Musquée; Rousselet Hâtif*).—Fruit small, pyriform. Skin smooth, yellow in the shade, and bright red next the sun, covered with grey dots. Eye small, placed in a shallow basin. Stalk an inch long, inserted without depression. Flesh yellowish, crisp, tender and juicy, sweet and perfumed.

An early pear, ripe in the end of July and beginning of August.

Early Sugar. See *Amiré Joannet.*

Easter Bergamot (*Bergamotte de Bugi; Bergamotte de Pâques; Bergamotte de Toulouse; Paddington; Roberts' Keeping; Royal Tairlon; Tarling*). — Fruit medium sized, roundish-turbinate. Skin pale green at first, but changing to pale yellow, and covered with numerous brownish-grey dots. Eye small, set in a shallow basin. Stalk an inch long, set in a small cavity. Flesh white, slightly gritty, crisp and juicy, sweet and aromatic. In use from March to April.

Easter Beurré (*Beurré de la Pentecôte; Beurré Anglaise; Beurré de Pâques; Beurré d'Hiver de Bruxelles; Bergamotte d'Hiver; Bergamotte de la Pentecôte; Bergamotte Tardive; Canning; Canning d'Hiver; Doyenné d'Hiver; Doyenné d'Hiver Nouveau; Doyenné de Pâques; Doyenné de Printemps; Du Pâtre; Philippe de Pâques; Seigneur d'Hiver; Sylvange d'Hiver*). — Fruit large, obovate, inclining to ovate. Skin pale green at first, but inclining to yellowish-green, and sometimes with a brownish tinge next the sun, marked with a few patches of russet, and strewed with numerous large russet dots. Eye small, with long, narrow, incurved segments set in a pretty deep cavity. Stalk an inch long, stout, inserted in a deep narrow cavity. Flesh white, buttery, and melting, very juicy, sugary, and richly flavoured.

One of the best late pears, in use from January till March.

Echassery (*Bezi d'Echassery; Bezi de Landry; Muscat de Villandry; Viandry; Verte Longue d'Hiver*). Fruit produced in clusters; medium sized, roundish-oval. Skin clear yellow, covered with numerous dots and patches of greyish-brown russet. Eye small and open, set in a shallow basin. Stalk an inch and a half long, inserted in a small knobbed cavity. Flesh white, buttery and melting, sugary, and with a musky flavour.

In use from November till Christmas.

Elisa d'Heyst.—Fruit above medium size, or large irregular-oval, widest in the middle and tapering towards the eye and the stalk. Skin smooth and shining, yellowish-green, clouded with russet about the stalk, and covered with russet dots. Eye closed, set in a deep, irregular basin. Stalk half an inch long, stout, and inserted without depression. Flesh melting, juicy, sugary, and richly flavoured. Ripe in February and March.

Ellanrioch. See *Hampden's Bergamot.*

Elton.—Fruit medium sized, oval. Skin greenish, almost entirely covered with thin grey russet, and marked with patches of coarser russet, with a tinge of orange on the part exposed to the sun. Eye small, very slightly depressed. Stalk stout, inserted in a deep cavity. Flesh firm, crisp, juicy, rich, and excellent. It is frequently without a core and pips, the flesh being solid throughout. Ripe in September, but does not keep long.

Emerald.—Fruit medium sized, obovate, rather uneven in its outline. Skin pale green, with pale brownish-red next the sun, and covered with russety dots. Eye open, set in a small irregular basin. Stalk an inch and a half long, obliquely inserted in a small cavity. Flesh buttery, melting, and richly flavoured. Ripe in November and December.

Emile d'Heyst.—Fruit above medium size, pyramidal. Skin bright yellow when ripe, marked with patches and veins of cinnamon-coloured russet. Eye small, set in a narrow and rather deep basin. Stalk about an inch long, set in a narrow uneven cavity. Flesh tender, buttery, and melting, very juicy, sugary, and perfumed. November.

English Bergamot. See *Autumn Bergamot.*

English Caillot Rosat. See *Caillot Rosat.*

Epargne. See *Jargonelle.*

Eparonanais. See *Duchesse d'Angoulême.*

Epine Dumas. See *Belle Epine du Mas.*

Epine d'Eté. See *Summer Thorn.*

Epine d'Eté Couleur de Rose. See *Summer Thorn.*

Epine d'Eté Vert. See *Summer Thorn.*

Epine d'Hiver. See *Winter Thorn.*

Epine de Rochechouart. See *Belle Epine du Mas.*

Epine Rose. See *Summer Rose.*

Epine Rose d'Hiver. See *Winter Thorn.*

Etourneau. See *Winter Nelis.*

Excellentissime. See *Fondante d'Automne.*

Eyewood. — Fruit below medium size, Bergamot-shaped. Skin greenish-yellow, very much covered with pale brown russet, and large russet dots. Eye small and open, slightly depressed. Stalk above an inch long, slender, inserted in a small cavity. Flesh yellowish, exceedingly tender and melting, very juicy, with a sprightly vinous flavour, and a fine aroma..

A very excellent pear, ripe in October. The tree is very hardy, and a good bearer.

Fanfareau. See *Hampden's Bergamot.*

Figue d'Alençon (*Bonnissime de la Sarthe; Figue d'Hiver.*—Fruit medium sized, pyriform. Skin greenish-yellow, strewed with russety dots. Eye small, set in a shallow basin. Stalk half an inch long, inserted obliquely, without depression. Flesh greenish, melting, juicy, sweet, and vinous. Ripe in November and December.

Figue d'Hiver. See *Figue d'Alençon.*

Figue Musquée. See *Windsor.*

Figue de Naples (*Comtesse de Frênol; Vigne de Pelone*). — Fruit above medium size, oblong. Skin greenish-yellow, entirely covered with thin, delicate russet, and dark reddish-brown on the side next the sun. Eye open, set in a wide, shallow basin. Stalk three quarters of an inch long, inserted without depression. Flesh greenish-white, buttery and melting, with a rich sugary flavour.

An excellent pear, ripe in November.

Fingal's. See *Hampden's Bergamot.*

Fin Or d'Eté. See *Summer Franc Real.*

Fin Or d'Hiver. See *Winter Franc Real.*

Flemish Beauty (*Belle de Flandre; Beurré des Bois;*

*Beurré de Bourgogne; Beurré Davy; Beurré Davis; Beurré d'Elberg; Beurré Foidard; Beurré St. Amour; Beurré Spence; Boss Père; Bosc Sire; Bouche Nouvelle; Brilliant; Fondante des Bois; Gagnée à Heuze; Impératrice des Bois.*)—Fruit large and obovate. Skin pale yellow, almost entirely covered with yellowish-brown russet on the shaded side, and reddish-brown on the side next the sun. Eye open, set in a small, shallow basin. Stalk an inch long, inserted in a rather deep cavity. Flesh yellowish-white, buttery and melting, rich and sugary. Ripe in September.

To have this excellent pear in perfection it should be gathered before it is thoroughly ripe, otherwise it is very inferior in quality.

FLEMISH BON CHRÊTIEN (*Bon Chrêtien Nouvelle; Bon Chrêtien Turc; Bon Chrêtien de Vernois*).—Fruit medium sized, obovate. Skin yellow, thickly strewed with russety dots, which are thickest on the side next the sun. Eye open, set in a small and shallow basin. Stalk an inch and a half long, inserted by the side of a fleshy swelling. Flesh yellowish-white, crisp, sweet, and perfumed.

An excellent stewing pear, in use from November till March.

Florence d'Eté. See *Summer Bon Chrêtien.*

FONDANTE D'AUTOMNE (*Arbre Superbe; Belle Lucrative; Bergamotte Fiévé; Beurré d'Albret; Beurré Lucratif; Excellentissime; Gresilière; Lucrate; Seigneur; Seigneur d'Esperen*).—Fruit large, obovate, and handsomely shaped. Skin lemon-yellow, with tinges of green over the surface, marked with patches of yellowish-brown russet. Eye small and open, set in a shallow basin. Stalk long, fleshy at the base, and obliquely inserted without depression. Flesh white, very tender, fine-grained and melting, very juicy, sugary, and aromatic.

A delicious autumn pear, ripe during September and October.

Fondante des Bois. See *Flemish Beauty.*

FONDANTE DES CHARNEUX (*Belle Excellente; Beurré des Charneuses; Desirée Van Mons; Duc de Brabant; Miel de Waterloo*).—Fruit large, pyriform, uneven in its outline. Skin greenish-yellow, with a faint tinge of red on the side next the sun, and thickly strewed with russet dots. Eye large and open, set in a shallow, uneven basin. Stalk upwards of an inch long, curved and inserted without depression by the side of a fleshy lip. Flesh tender, buttery, and melting, sugary, and richly flavoured. Ripe in November.

Fondante de Jaffard. See *Colmar d'Aremberg.*

Fondante de Malines. See *Winter Nelis.*

Fondante de Mons. See *Passe Colmar.*

Fondante Musquée. See *Summer Thorn.*

Fondante de Noël (*Belle après Noël; Belle de Noël; Bonne de Noël*).—Fruit medium sized, turbinate. Skin yellow next the sun, covered with traces of russet and numerous russet dots, sometimes tinged with red on the side. Eye closed, set in a broad, shallow basin. Stalk long, obliquely inserted by the side of a fleshy lip. Flesh melting, juicy, sweet, and well flavoured. December and January.

Fondante de Parisel. See *Passe Colmar.*

Fondante Van Mons.—Fruit medium sized, roundish, and somewhat depressed. Skin thin and delicate, of a fine waxen-yellow colour, mottled with very thin cinnamon-coloured russet. Eye open, set in a very shallow depression. Stalk an inch long, set in a narrow and rather deep cavity. Flesh white, juicy, melting, and sugary, with a slightly perfumed flavour. September and October.

An excellent dessert pear, but not so rich as Fondante d'Automne, which is ripe at the same time.

Forelle (*Trout; Truit*).—Fruit medium sized, oblong-obovate, but sometimes assuming a pyriform shape. Skin smooth and shining, of a fine lemon-yellow colour on the shaded side, and bright crimson on the side next the sun, covered with numerous crimson spots, which from their resemblance to the markings on a trout have suggested the name. Eye small, set in a rather shallow basin. Stalk an inch long, slender, inserted in a small shallow cavity. Flesh white, delicate, buttery, and melting, with a rich sugary and vinous flavour.

An excellent pear, in use from November till February. The tree is hardy, and a good bearer.

Fortunée (*Bergamotte Fortunée; Fortunée Parmentier*).—Fruit below medium size, roundish-turbinate, uneven in its outline. Skin deep yellow, covered all over with flakes and lines of brown russet. Eye closed, deeply sunk. Stalk three quarters of an inch long, stout. Flesh half-melting, juicy, and sweet.

A stewing pear, in use from January till May.

De Fosse. See *Jargonelle.*

Franc Real d'Eté. See *Summer Franc Real.*

Franc Real Gros. See *Angélique de Bordeaux.*

Franc Real d'Hiver. See *Winter Franc Real.*

Frederic Le Clerc.—Fruit above medium size, short, pyriform. Skin green at first, but changing as it ripens to yellow; slightly mottled with russet. Eye open, set in a shallow basin. Stalk an inch long, woody. Flesh yellowish, buttery, melting, and very juicy, sugary, and rich. Ripe in November.

Frederic de Wurtemburg. — Fruit large, obtuse-pyriform. Skin smooth, deep yellow, marbled and dotted with red on the shaded side, and of a beautiful bright crimson next the sun. Eye large and open, placed almost level with the surface. Stalk thick. Flesh very white, tender, buttery, and melting, rich, juicy, sugary, and delicious.

A remarkably fine pear, ripe in October. In the year 1858 it was as finely flavoured as the Jargonelle.

Gagnée à Heuze. See *Flemish Beauty.*

Galston Moorfowl's Egg. — Fruit below medium size, short obovate. Skin greenish-yellow, entirely covered with thin pale brown russet, and mottled with red next the sun. Eye open, set in a wide, shallow basin. Stalk about an inch long. Flesh yellowish, tender, sweet, and juicy.

An excellent Scotch pear with a peculiar aroma, ripe in the end of September.

Gambier. See *Passe Colmar.*

Gansel's Bergamot (*Bonne Rouge; Brocas' Bergamot; Diamant; Gurle's Beurré; Ive's Bergamot; Staunton*).— Fruit above medium size, or large; roundish-obovate, and flattened at the apex. Skin greenish-yellow on the shaded side, and reddish-brown next the sun, the whole thickly strewed with russety dots and specks. Eye small and open, set in a shallow basin. Stalk short and fleshy. Flesh white, buttery, melting, and very juicy, sugary and aromatic.

A fine old dessert pear, ripe during October and November. In warm situations it ripens well on a standard, but it generally requires a wall.

Garde Ecosse. See *Gilogil.*

GENDESHEIM (*Verlaine; Verlaine d'Eté*).—Fruit medium sized, obtuse-pyriform. Skin pale greenish-yellow, thickly covered with grey russety dots. Eye small and open, placed in a shallow depression. Stalk an inch long, inserted in a small cavity. Flesh buttery, with a rich sugary and somewhat musky flavour.

An excellent pear, in use during October and November.

GENERAL TODTLEBEN.—Fruit very large, pyriform. Skin yellow, covered with dots and patches of brown russet. Eye open, set in a wide furrowed basin. Stalk an inch long, set in a small cavity. Flesh with a rosy tinge, very melting and juicy, slightly gritty, with a rich, sugary, and perfumed juice.

A new Belgian pear, which fruited for the first time in 1855, said to be very excellent. In use from December to February.

German Baker. See *Uvedale's St. Germain.*

GILOGIL (*Beurré Geerards; Garde Ecosse; Gil-ô-gile; Gobert; Gros Gobet*).—Fruit very large, roundish-turbinate. Skin yellowish in the shade and brownish next the sun, entirely covered with thin brown russet. Eye large, set in a deep and plaited basin. Stalk an inch long, deeply inserted. Flesh firm, crisp, sweet, and juicy.

An excellent stewing pear, in use from November to February.

De Glace. See *Virgouleuse.*

GLOU MORCEAU (*Beurré de Cambronne; Beurré d'Hardenpont; Beurré de Kent; Beurré Lombard; De Cambron; Colmar d'Hiver; Got Luc de Cambron; Goulu Morceau; Hardenpont d'Hiver; Linden d'Automne; Roi de Wurtemburg*).—Fruit above medium size, obovate, narrowing obtusely from the bulge to the eye and the stalk. Skin smooth, pale greenish-yellow, covered with greenish-grey russet dots, and slight markings of russet. Eye open, set in a rather deep basin. Stalk an inch and a half long, inserted in a narrow cavity. Flesh white, tender, smooth, and buttery, of a rich and sugary flavour.

A first-rate dessert pear, in use from December to January.

Gobert. See *Gilogil.*

GOLDEN KNAP.—This is a very small roundish-turbinate russety pear, of no great merit. It is grown extensively in the orchards of the border counties and in the Carse of Gowrie; and, being a prodigious and constant bearer, is well adapted for orchard planting where quantity and not quality is the object. Ripe in October.

Got Luc de Cambron. See *Glou Morceau.*

Goulu Morceau. See *Glou Morceau.*

Gracieuse. See *Belle et Bonne.*

Grand Monarque. See *Catillac.*

GRAND SOLEIL.—Fruit large, roundish-turbinate. Skin very rough to the feel, entirely covered with dark-brown russet of the colour of that which covers the Royal Russet apple. Eye open, set in a pretty deep basin. Stalk an inch and a quarter long, thick and fleshy, swelling out at the base into the substance of the fruit. Flesh white, coarse-grained, crisp and very juicy, sweet and sugary, with a pleasant flavour. November.

Gratioli. See *Summer Bon Chrêtien.*

Gratioli d'Hiver. See *Beurré Diel.*

Gratioli di Roma. See *Summer Bon Chrêtien.*

Great Bergamot. See *Hampden's Bergamot.*

GREEN CHISEL.—Fruit very small, growing in clusters, roundish-turbinate. Skin green, with sometimes a brownish tinge next the sun. Eye large and open. Stalk three quarters of an inch long, inserted without depression. Flesh juicy and sweet.

An old-fashioned early pear, of little merit. Ripe in August.

Green Windsor. See *Windsor.*

GREEN YAIR.—Fruit below medium size, obovate. Skin smooth, dark green, changing to yellowish-green as it ripens, and strewed with patches and dots of russet.

Eye large, open, and prominent. Stalk three quarters of an inch long, obliquely inserted. Flesh tender, juicy, and sugary.

A good Scotch pear, ripe in September.

Gresilière. See *Fondante d'Automne.*

Grey Achan. See *Chaumontel.*

Grey Doyenné. See *Red Doyenné.*

Grey Goose. See *Gros Rousselet.*

Groom's Princess Royal. — Fruit medium sized, roundish. Skin greenish, marked with russet, and with a brownish tinge next the sun. Eye small and open, set in a slight depression. Stalk short and thick. Flesh buttery, melting, rich, and sugary. In use from January till March.

Gros Gilot. See *Catillac.*

Gros Gobet. See *Gilogil.*

Gros Micet. See *Winter Franc Real.*

Gros Rousselet (*Gros Rousselet de Rheims; Grey Goose; Roi d'Eté*).—Fruit medium sized, obtuse-pyriform, and rounded at the apex. Skin of a fine deep yellow colour, with brownish-red next the sun, and thickly strewed with russety dots. Eye small and open. Stalk an inch and a half to two inches long. Flesh white, tender, half-melting, very juicy, vinous, and musky. August and September.

Gros Rousselet de Rheims. See *Gros Rousselet.*

Gros St. Jean. See *Citron des Carmes.*

Grosse Cuisse Madame. See *Jargonelle.*

Grosse Dorothée. See *Beurré Diel.*

Grosse Jargonelle. See *Windsor.*

Grosse Ognonet. See *Summer Archduke.*

Guernsey Chaumontel. See *Chaumontel.*

Gurle's Beurré. See *Gansel's Bergamot.*

Hacon's Incomparable (*Downham Seedling*).—Fruit above medium size, roundish. Skin pale yellowish-green, sometimes with a brownish tinge on one side, and strewed with russety dots. Eye small and open, set in a shallow basin. Stalk an inch long. Flesh white, buttery and melting, with a rich sugary, vinous, and highly perfumed flavour.

An excellent hardy pear, in use from November to January.

Hampden's Bergamot (*Belle d'Août; Belle de Bruxelles; Belle sans Epines; Bergamotte d'Eté Grosse; Bergamotte de Paysans; Ellanrioch; Fanfareau; Fingaïs; Great Bergamot; Longueville; Scotch Bergamot*). — Fruit above medium size, abrupt pyriform. Skin smooth, of a fine clear lemon yellow, strewed with dots and flakes of thin pale brown russet, and with a tinge of bright red on the side next the sun. Eye rather small, set in an uneven shallow basin. Stalk an inch long, inserted without depression. Flesh pure white, tender, melting, and juicy, sweet, and with a high aroma.

A fine showy and excellent early pear, ripe in the middle and end of August, but soon decays at the core.

Hardenpont d'Hiver. See *Glou Morceau.*

Hardenpont de Printemps. See *Beurré de Rance.*

Harvest Pear. See *Amiré Joannet.*

Hazel. See *Hessle.*

Heliote Dundas (*Rousselet Jamain*).—Fruit medium sized, pyriform, even, and regularly formed. Skin smooth and somewhat shining, lemon yellow, with a brilliant red cheek, dotted with large dark-red specks. Eye small, and deeply set. Stalk upwards of an inch long. Flesh white, half-buttery, and not very juicy; very sweet, piquant, and perfumed. Ripe in October, and soon rots at the core.

Henri Capron.—Fruit medium sized, egg-shaped. Skin pale yellow, mottled with pale brown, sprinkled with flakes and dots of delicate russet. Eye nearly closed. Stalk three quarters of an inch long, stout. Flesh yellowish-white, buttery, and highly aromatic. Ripe in October and November.

Henri Quatre. See *Henry the Fourth.*

Henriette Bouvier. — Fruit about medium size, roundish-obovate. Skin pale yellow, covered with patches and network of smooth cinnamon-coloured russet, and sometimes with an orange tinge next the sun. Eye small, and almost level with the surface. Stalk an inch or more in length, inserted without depression. Flesh very tender, buttery and melting, very rich and sugary, with a fine perfume.

A very fine pear, ripe in the beginning and middle of December.

Henry the Fourth (*Henri Quatre; Jacquin*).—Fruit small, obtuse-pyriform. Skin greenish, pale yellow, considerably covered with pale cinnamon-coloured russet, and grey specks. Eye small and open. Stalk an inch long, obliquely inserted. Flesh white, rather coarse-grained, but very juicy and melting, with a rich, sugary, and aromatic flavour.

A most delicious little pear, ripe in September.

Hessle (*Hazel; Hessel*).—Fruit below medium size, turbinate. Skin greenish-yellow, very much covered with large russety dots, which give it a freckled appearance. Eye small and open, slightly depressed. Stalk an inch long, obliquely inserted without depression. Flesh tender, very juicy, sweet, and with a high aroma.

An excellent market-gardening pear, ripe in October. The tree is a most abundant and regular bearer.

His. See *Baronne de Mello.*

Holland Bergamot. See *Bergamotte d'Hollande.*

Hubard. See *Beurré d'Amanlis.*

Huntingdon. See *Lammas.*

Huyshe's Bergamot—Fruit large, inclining to obovate. Skin tolerably smooth, considerably covered with russet. Eye somewhat open, moderately depressed. Stalk short, thick, and obliquely inserted in a narrow cavity. Flesh yellowish-white, exceedingly melting and juicy, somewhat gritty at the core; rich, sugary, and delicious.

A remarkably fine pear, in use in the end of December and January.

Huyshe's Victoria.—Fruit medium sized, oval and almost cylindrical, flat at the ends. Skin yellowish, freckled with russet. Eye small, set in a shallow depression. Stalk very short and thick, not deeply inserted. Flesh melting, rather gritty at the core, juicy, rich, and sugary, with a brisk acidity. In use during December and January, but not equal to the preceding.

Imperatrice de Bois. See *Flemish Beauty.*

L'Inconnue (*L'Inconnue Van Mons*).—Fruit large and pyriform. Skin rough to the feel, greenish-yellow, covered with large grey dots and patches of cinnamon-coloured russet. Eye small and sometimes wanting, set in a deep basin. Stalk an inch to an inch and a quarter long, inserted without depression. Flesh yellowish, firm, very juicy, rich, and sugary, with an agreeable aroma.

A very excellent winter pear, ripe in February.

Inconnue la Fare. See *St. Germain*.

Isambert le Bon. See *Brown Beurré*.

Ive's Bergamot. See *Gansel's Bergamot*.

Jackman's Melting. See *King Edward's*.

Jacquin. See *Henry the Fourth*.

Jalousie de Fontenay (*Belle d'Esquerme; Jalousie de Fontenay Vendée*).—Fruit medium sized, obtuse-pyriform. Skin greenish-yellow, tinged with red on the exposed side, and covered with russety dots and patches. Eye closed. Stalk an inch long. Flesh white, buttery, melting, and richly flavoured. October and November.

Jaminette (*Austrasie; Banneaux; Belle d'Austrasie; Bergamotte d'Austrasie; Colmar Jaminette; Crassanne d'Austrasie; Josephine; Maroit; Pyrole; Sabine*).—Fruit medium sized, turbinate. Skin pale yellowish-green, thickly covered with brown dots, and marked with cinnamon-coloured russet next the sun and round the stalk. Eye open, set in a rather deep basin. Stalk about an inch long, obliquely inserted. Flesh white, very juicy and melting, sugary and vinous.

A first-rate pear, in use from November to January.

Jargonelle (*Beau Présent; Belle Vièrge; Beurré de Paris; Chopine; Cueillette; Epargne; De Fosse; Grosse Cuisse Madame; Mouille Bouche d'Eté; Sweet Summer; St. Lambert; St. Samson; De la Table des Princes*).—Fruit large and pyriform. Skin smooth, greenish-yellow, with a tinge of dark brownish-red next the sun. Eye large and open. Stalk about two inches long, slender, and obliquely inserted without depression. Flesh yellowish-white, tender, melting, and very juicy, with a rich piquant flavour, and slight musky aroma.

A first-rate pear, ripe in August.

Jean de Witte.—Fruit medium sized, obovate. Skin smooth, of a greenish-yellow colour, covered with numerous small grey dots, and a few markings of thin cinnamon-coloured russet. Eye small and closed, rather deeply set. Stalk an inch or more in length. Flesh yellowish, fine-grained, buttery, and melting, with a rich sugary flavour not unlike that of Glou Morceau.

A first-rate pear, in use from January till March.

Jersey Gratioli (*Bedminster Gratioli; Norris' Pear*).—Fruit above medium size, roundish-obovate. Skin greenish-yellow, covered with large, rough, russet spots, and tinged with pale brown next the sun. Eye open, set in an even, shallow basin. Stalk an inch long, in a narrow cavity. Flesh yellowish-white, very melting, rich, sugary, and with a fine sprightly vinous flavour.

A very excellent pear. Ripe in October.

Jewess (*La Juive*).—Fruit medium sized, pyramidal. Skin of a uniform pale yellow colour, mottled with pale brown russet, and thickly covered with russet dots. Eye small and open, with short, erect segments even with the surface. Stalk about an inch long, stout, and tapering into the fruit, or obliquely inserted. Flesh yellowish, buttery, and melting, very juicy, sugary, and rich.

A most delicious pear. Ripe in December.

Joannet. See *Amiré Joannet*.

John. See *Monsieur Jean*.

John Dory. See *Monsieur Jean*.

Joséphine. See *Jaminette*.

Joséphine de Malines.—Fruit about medium size. Skin yellow, with a greenish tinge on the shaded side and with a tinge of red on the side next the sun; the whole surface strewed with large russet spots. Eye open, set in a rather shallow depression. Stalk three quarters of an inch long, stout, and inserted in a narrow cavity. Flesh yellowish, with a tinge of red, melting and very juicy, sugary, vinous, and richly flavoured, with a high rosewater aroma.

A most delicious pear, in use from February till May. The tree is hardy, and an excellent bearer.

La Juive. See *Jewess*.

Kaissoise. See *Beurré d'Amanlis*.

Kartofell. See *Colmar d'Aremberg*.

De Kienzheim. See *Vallée Franche*.

King Pear. See *Caillot Rosat*.

KING EDWARD'S (*Jackman's Melting*).—Fruit very large, the size and shape of Uvedale's St. Germain. Skin smooth and shining, of a beautiful grass-green colour, which it retains even when ripe, and with a flush of reddish-brown on the exposed side, thickly dotted all over with large brown russet dots. Eye open, set in a narrow, plaited basin. Stalk an inch long, inserted without depression. Flesh fine-grained, tender and melting, juicy, but not very sugary, and with a perfume of musk.

The largest really melting pear, and, for its size, very good. Ripe in September and October.

Knight's Monarch. See *Monarch*.

Konge. See *Windsor*.

Lafare. See *St. Germain*.

LAMMAS (*Huntingdon*).—Fruit below medium size, pyramidal, regular and handsome. Skin pale yellow, streaked with red, and covered with red on the side next the sun. Eye open, very slightly depressed. Stalk half an inch long, inserted without depression. Flesh tender, juicy, and melting, with an agreeable flavour. Ripe in the beginning and middle of August.

The tree is hardy, and a most abundant bearer.

Lammas [of the Scotch]. See *Crawford*.

LAURE DE GLYMES.—Fruit above medium size, pyramidal. Skin entirely covered with a coat of fawn-coloured russet, with mottles of lemon-coloured ground shining through. Eye open, set in a shallow basin. Stalk an inch long, stout and fleshy, not depressed. Flesh white, tender and juicy, sweet and highly perfumed. Ripe in the beginning of October.

De Lavault. See *Williams' Bon Chrétien*.

Lent St. Germain. See *Uvedale's St. Germain*.

LÉON LE CLERC DE LAVAL.—Fruit large, long-obovate, and rounding towards the eye. Skin smooth and shining, yellow, strewed with brown dots, and marked with tracings of russet. Eye large, with long, straight, narrow segments, set in a shallow basin. Stalk an inch and a half long, inserted without depression by the side of a fleshy lip. Flesh white, half-melting or crisp, juicy, sweet, and perfumed.

An excellent stewing pear, which in some seasons is

half-melting, and is in use from January till May and June.

Léon le Clerc de Louvain.—Fruit medium sized, longish-oval, and blunt at both ends. Skin of a yellow colour, washed with red on the side next the sun. Eye large and closed. Stalk an inch long, and pretty thick. Flesh yellowish, half-melting, juicy, sweet, and pretty well flavoured. Ripe in the middle of November.

Both of the above are very distinct pears from Van Mons Léon le Clerc.

Leopold the First.—Fruit medium sized, oval, inclining to pyriform. Skin greenish-yellow, covered with flakes and dots of russet. Eye open, irregular, slightly depressed. Stalk an inch long, thick and curved. Flesh yellowish-white, melting, very juicy, rich, sugary, and highly perfumed.

A first-rate dessert pear. Ripe in December and January. The tree forms a very handsome pyramid.

Lewis.—Fruit medium sized, oblong-obovate. Skin pale green, assuming a yellow tinge as it ripens, thickly covered with brown russet dots and with patches of russet round the stalk and the eye. Eye large and open, slightly depressed. Stalk an inch and three quarters long, slender, and inserted without depression. Flesh yellowish-white, very tender, melting, and very juicy, rich and sugary, with a somewhat aromatic flavour.

An excellent pear, in use from November to January. The tree is an abundant bearer and hardy.

Liard. See *Napoléon*.

Linden d'Automne. See *Glou Morceau*.

Lodge.—Fruit about medium size, obtuse-pyriform. Skin smooth and shining, yellowish-green, mottled with darker green; marked with a few flesh-coloured dots on the side next the sun, and strewed all over with faint tracings of delicate russet. Eye closed, set in a shallow basin. Stalk upwards of an inch long, slender, inserted without depression. Flesh white, tender, melting, and juicy, but with no particular aroma or flavour. Ripe in October.

This is somewhat like Louise Bonne of Jersey, but very inferior to that variety.

Longueville. See *Hampden's Bergamot*.

London Sugar.—Fruit below medium size, turbinate. Skin pale green, becoming yellow when ripe, with a brownish tinge when fully exposed to the sun. Eye small, half-open, prominent, and surrounded with puckered plaits. Stalk an inch long, slender, obliquely inserted. Flesh tender, melting, very juicy, sugary, and musky. Ripe in the end of July and beginning of August.

Lord Cheyne's. See *Bergamotte d'Hollande.*

Louise d'Avranches. See *Louise Bonne of Jersey.*

Louise Bonne of Jersey (*Beurré d'Avranches; Bonne d'Avranches; Bonne de Longueval; Bonne Louise d'Arandoré; Louise d'Avranches; William the Fourth*).—Fruit medium sized, pyriform. Skin smooth, yellow on the shaded side, but crimson next the sun, covered with crimson and russety dots. Eye small and open, set in a rather deep basin. Stalk three quarters of an inch long, obliquely inserted without depression. Flesh white, buttery, and melting, with a rich, sugary, and brisk vinous flavour.

A most delicious pear, ripe in October. The tree is a good bearer, and succeeds well as a pyramid on the quince.

Lucrate. See *Fondante d'Automne.*

Mabile. See *Napoléon.*

Madame. See *Windsor.*

Madame Durieux.—Fruit medium sized, bergamot-shaped. Skin greenish-yellow, mottled with large patches of russet, particularly about the stalk, and dotted and streaked with the same. Eye closed, slightly depressed. Stalk three quarters of an inch long. Flesh white, melting, buttery, juicy, and with a bergamot flavour. Ripe in the end of October and beginning of November.

Madame de France. See *Windsor.*

Madeleine. See *Citron des Carmes.*

March Bergamot.— Fruit medium sized, bergamot-shaped. Skin yellow, covered with minute russet dots, which cause it to feel rough. Eye open, set in a wide, even basin. Stalk an inch or more long, woody, inserted in a deep, round cavity. Flesh yellowish, firm, breaking half-melting, very juicy, and with a high bergamot flavour,

An excellent pear for the season. Ripe during March. and April.

Marianne Nouvelle. See *Beurré Bosc.*

Marie Chrêtienne. See *Marie Louise.*

Marie Louise (*Braddick's Field Standard; Marie Chrêtienne; Marie Louise Delcourt; Princesse de Parme; Van Doncklelaar*).—Fruit large, oblong or pyriform. Skin smooth, greenish-yellow, marked with tracings of thin brown russet. Eye small and open, set in a narrow, rather deep and uneven basin. Stalk an inch and a half long, inserted obliquely without depression. Flesh white, delicate, buttery, and melting, very juicy, and exceedingly rich, sugary, and vinous.

One of our very best pears. Ripe in October and November. The tree is an excellent bearer; but the bloom is tender. It succeeds well either on the pear or the quince, forming a handsome pyramid.

Marie Louise Delcourt. See *Marie Louise.*

Marie Louise Nova. See *Comte de Lamy.*

Maréchal de la Cour. See *Conseiller de la Cour.*

Maroit. See *Jaminette.*

Marotte Sucré. See *Passe Colmar.*

Martin Sec (*Dry Martin; Martin Sec de Champagne; Martin Sec d'Hiver*).—Fruit medium sized, obtuse-pyriform. Skin smooth and delicate, entirely covered with cinnamon-coloured russet on the shaded side, and bright red next the sun. Eye small and open, set in a plaited basin. Stalk an inch and a half long, inserted in a small cavity. Flesh breaking, rather dry, but sweet and perfumed.

An excellent stewing pear, in use from November till January.

Martin Sec de Champagne. See *Martin Sec.*

Martin Sec d'Hiver. See *Martin Sec.*

De Maune. See *Colmar.*

Medaille. See *Napoléon.*

Melon. See *Beurré Diel.*

Messire Jean (*Chaulis; John; John Dory; Messire Jean Blanc; Messire Jaune Doré; Monsieur John*).—Fruit medium-sized, turbinate, inclining to obovate. Skin greenish-yellow, thickly covered with brown russet. Eye small and open. Stalk an inch and a half long. Flesh white, crisp, juicy, sugary, and gritty.

A dessert pear of little merit. Ripe in November and December.

Miel de Waterloo. See *Fondante Charneux.*

Milanaise Cuvelier. See *Winter Nelis.*

Millet de Nancy.—Fruit rather below medium size, pyriform. Skin smooth, light green, becoming yellow at maturity. Flesh pale yellow, buttery, melting, and juicy, sugary, and agreeably perfumed. Ripe in October and November.

Moccas.—Fruit medium sized, oval, uneven and bossed in its outline. Skin lemon coloured, marked with patches and veins of thin pale brown russet, and strewed with russet dots. Eye somewhat closed, set in a deep, uneven, and furrowed basin. Stalk an inch long, rather deeply inserted. Flesh yellowish, fine-grained, tender and melting, with a rich vinous juice and musky flavour.

A very fine pear. Ripe in December and January.

Monarch (*Knight's Monarch*).—Fruit medium sized, roundish. Skin yellowish-green, very much covered with brown russet, and strewed with grey-russet specks. Eye small and open, set in a shallow undulating basin. Stalk three quarters of an inch long, inserted in a small cavity, frequently without depression. Flesh yellowish, buttery, melting, and very juicy, with a rich, piquant, sugary, and agreeably perfumed flavour.

One of the most valuable pears. Ripe in December and January. The tree is very hardy, an excellent bearer, and forms a handsome pyramid.

Monsieur de Clion. See *Vicar of Winkfield.*

Monsieur le Curé. See *Vicar of Winkfield.*

Monsieur John. See *Messire Jean.*

Monstrueuse de Landes. See *Catillac.*

Morel.—Fruit about medium sized, obovate. Skin yellow, thickly freckled with large russet spots. Eye half open, not depressed. Stalk an inch and a quarter long, stout. Flesh yellowish-white, crisp, juicy, and sweet, with an agreeable flavour.

This in colour and flavour is like Hessle, but ripens in April, and is a good variety for that late season.

Mouille Bouche. See *Verte Longue.*

Mouille Bouche d'Automne. See *Verte Longue.*

Mouille Bouche d'Eté. See *Jargonelle.*

MUIRFOWL'S EGG.—Fruit below medium size, roundish. Skin entirely covered with fine cinnamon-coloured russet, brownish-red next the sun, and thickly covered with grey-russet dots. Eye half open, set in a round depression. Stalk an inch long, set in a small, round cavity. Flesh tender, juicy, sweet, and brisk, with a strong musky perfume. Ripe in October.

Muscat de Villandry. See *Echassery.*

NAPOLÉON (*Bonaparte; Bon Chrêtien Napoléon; Beurré Napoléon; Captif de St. Hélène; Charles X., Gloire de l'Empereur; Liard; Mabile; Medaille; Napoléon d'Hiver; Roi de Rome; Sucrée Doré; Wurtemburg*).—Fruit large, obtuse-pyriform. Skin smooth, greenish-yellow, covered with numerous brown dots. Eye partially open, moderately depressed. Stalk three quarters of an inch long, stout, and inserted in a round, pretty deep, cavity. Flesh white, tender, melting, and very juicy, with a rich, sugary, and refreshing flavour.

A first-rate pear. Ripe in November and December. Succeeds best against a wall.

NAVEZ PEINTRE.—Fruit medium sized, egg-shaped, even and regularly formed. Skin yellowish-green on the shaded side, and marked with bands of brown russet, but with a blush of brownish-red next the sun. Eye open, very slightly depressed. Stalk an inch long, rather slender, not depressed. Flesh yellowish, melting very juicy, piquant, and sugary, with a fine aroma.

A very fine pear. Ripe in the end of September.

Neige. See *White Doyenné.*

Neige Grise. See *Red Doyenné.*

Nelis d'Hiver. See *Winter Nelis.*

NE PLUS MEURIS.—Fruit medium sized, roundish-turbinate, very uneven, and bossed on its surface. Skin rough, dull yellow, very much covered with dark brown russet. Eye half open, generally prominent. Stalk very short, not at all depressed, frequently appearing as a mere knob on the apex of the fruit. Flesh yellowish-white, buttery and melting, with a rich, sugary, and vinous flavour.

A first-rate pear. Ripe from January till March. It succeeds well as a pyramid, but is best from a wall.

Ne Plus Meuris [of the French]. See *Beurré d'Anjou.*

Neuve Maisons.—Fruit large, pyramidal, even and regularly formed. Skin smooth, of a uniform yellow colour, thickly strewed with large russet dots, and a few patches of thin russet. Eye open, set in a narrow and round basin. Stalk an inch or more in length, very stout, inserted in a narrow depression. Flesh coarse-grained, melting, with a thin, somewhat vinous, juice, but without much flavour. Ripe in October and November.

New Autumn. See *Jargonelle.*

New York Red-Cheek. See *Seckle.*

Notaire Minot. — Fruit medium sized, roundish-obovate. Skin pale yellowish-green, considerably covered with patches and large dots of rough brown russet. Eye open, set in a narrow and shallow basin. Stalk an inch long, stout, inserted by the side of a fleshy lip. Flesh yellowish, rather coarse-grained, but melting, and with a fine brisk, vinous, and sugary flavour.

A very good pear. Ripe in January and February.

Nouveau Poiteau (*Tombe de l'Amateur*). — Fruit very large, obtuse-obovate or pyramidal. Skin greenish-yellow, or pale yellow, mottled and streaked with pale brown russet. Eye closed, placed in a slight depression. Stalk an inch to an inch and a quarter long, obliquely inserted in a small cavity. Flesh fine-grained, buttery, melting, and very juicy, rich, sugary, and highly perfumed.

A first-rate pear. Ripe during November, but keeps only a short time.

Nouvelle Boussoch. See *Doyenné Boussoch.*

Nutmeg. See *Bezi de Caissoy.*

Œuf. — Fruit small, oval. Skin smooth, greenish-yellow, marked with light red on the exposed side, and strewed with grey russety dots. Eye small and open, set in an uneven depression. Stalk an inch long, inserted in a small cavity. Flesh whitish, tender and melting, rich, sugary, and musky.

A very good summer pear. Ripe in August, and keeps for three weeks without decaying, which is a recommendation at this season.

Ognonet. See *Summer Archduke.*

Ognonet Musqué. See *Summer Archduke.*

Oken d'Hiver. See *Winter Oken.*

L'Orpheline. See *Beurré d'Aremberg.*

L'Orpheline d'Enghein. See *Beurré d'Aremberg.*

Oxford Chaumontel. See *Chaumontel.*

Paddington. See *Easter Bergamot.*

Paradise d'Automne.—Fruit below medium size, pyriform. Skin covered with a coat of rough, dark cinnamon coloured russet, which is strewed with grey dots. Eye very small and open, set in a shallow basin. Stalk an inch and a quarter long, obliquely inserted without depression. Flesh fine-grained, buttery and melting, rich, sugary, and with a fine piquant and perfumed flavour.

A remarkably fine pear. Ripe in October and November.

Parkinson's Warden. See *Black Worcester.*

Passans de Portugal.—Fruit medium sized, oblate. Skin pale yellow, with a lively red cheek. Eye open, set in a shallow depression. Stalk an inch long, inserted in a small round cavity. Flesh white, crisp, juicy, sugary, and perfumed. Ripe in the end of August and beginning of September.

Passe Colmar (*Beurré d'Argenson; Cellite; Chapman's; Chapman's Passe Colmar; Colmar Doré; Colmar Epineux; Colmar d'Hardenpont; Colmar Preul; Colmar Souveraine; Fondante de Parisel; Fondante de Mons; Gambier; Marotte Sucré; Passe Colmar Doré; Passe Colmar Epineux; Passe Colmar Gris; Precel; Présent de Malines; Pucelle Condesienne; Regentin; Souverain*).—Fruit medium sized, obovate. Skin smooth, of a fine uniform deep lemon colour, with a tinge of red on the side next the sun, strewed with numerous brown dots and veins of russet. Eye open, set in a wide shallow basin. Stalk from three quarters to an inch long, inserted in a small sheath-like cavity. Flesh yellowish-white, buttery, melting, and very juicy, with a rich, sugary, vinous, and aromatic flavour.

An excellent pear. Ripe during November and December. The tree is an excellent bearer, and forms a handsome pyramid. It requires a rich, warm soil, otherwise the flesh is crisp and gritty. In exposed situations it requires a wall.

Passe Colmar Doré. See *Passe Colmar.*

Passe Colmar Epineux. See *Passe Colmar.*

Passe Colmar Gris. See *Passe Colmar.*

Passe Madeleine.—This is a small oblong pear with an uneven surface. Skin green, covered with dots. The flesh is dry and very astringent, crisp and without much flavour.

An early pear. Ripe in August, and grown to some extent in the market-gardens round London; but it is a very worthless variety.

Paternoster. See *Vicar of Winkfield.*

Du Patre. See *Easter Beurré.*

Peach (*Pêche*).—Fruit medium sized or large; irregularly oval or roundish. Skin smooth, greenish-yellow, with a blush of red on the side next the sun, and covered with patches and dots of russet. Eye open, set in a shallow bossed basin. Stalk an inch or more long, not depressed. Flesh yellowish-white, fine-grained, and very melting, very juicy, sugary, vinous, and with a delicious perfume.

An excellent early pear. Ripe in the middle and end of August.

Pêche. See *Peach.*

Pengthley.—Fruit medium sized, obovate, inclining to oval. Skin pale green, covered with dark dots, and becoming yellow as it ripens. Eye large and open, set in a shallow depression. Stalk long and slender, curved, and set in an uneven cavity. Flesh coarse-grained, crisp, very juicy and sweet. Ripe in March.

Perdreau. See *Early Rousselet.*

Perdreau Musqué. See *Early Rousselet.*

Petit Beurré d'Hiver. See *Bezi de Caissoy.*

Petit Muscat (*Little Muscat; Sept-en-gueule*).—Fruit very small, produced in clusters, turbinate. Skin bright yellow when ripe, and covered with brownish-red next the sun, and strewed with russet dots. Eye open, not depressed. Stalk about an inch long, not depressed. Flesh melting, sweet, juicy, and with a musky flavour.

A very early pear. Ripe in the end of July.

Petit St. Jean. See *Amiré Joannet.*

De Pézénas. See *Duchesse d'Angoulême.*

Philippe Delfosse. See *Beurré Delfosse.*

Philippe de Pâques. See *Easter Beurré.*

Pickering Pear. See *Uvedale's St. Germain.*

Pickering's Warden. See *Uvedale's St. Germain.*

Pine. See *White Doyenné.*

Piper. See *Uvedale's St. Germain.*

Piquery. See *Urbaniste.*

Pitt's Calabasse. See *Calebasse.*

PIUS IX.—Fruit large, conical, and regularly formed. Skin of a deep, clear yellow colour, with a blush of red on the side next the sun, considerably covered with streaks and flakes of russet. Eye open, slightly depressed. Stalk thick and woody, very short. Flesh melting, juicy, sugary, and highly perfumed.

An excellent pear. Ripe in September. The tree is hardy, of small habit, forms a nice pyramid, and is a good bearer.

Plombgastelle. See *Beurré d'Amanlis.*

Poire de Prince. See *Chair à Dames.*

Pound Pear. See *Black Worcester.*

Pound Pear. See *Catillac.*

Precel. See *Passe Colmar.*

Présent de Malines. See *Passe Colmar.*

Présent Royal de Naples. See *Beau Présent d'Artois.*

PRÉVOST.—Fruit rather large, roundish-oval. Skin clear golden yellow, with a bright red blush on the exposed side, and marked with flakes of russet. Eye open, not deeply sunk. Stalk about an inch long. Flesh fine-grained, half-melting, and half-buttery, pretty juicy, and highly aromatic.

A good late pear. Ripe from January to April; but unless grown in a warm soil and situation it rarely attains the character of a melting pear.

PRINCE ALBERT.—Fruit medium sized, pyriform. Skin smooth, of a deep lemon-yellow colour, and frequently with a blush of red next the sun. Eye small and open, set in a shallow basin. Stalk an inch long, not depressed. Flesh yellowish-white, melting, juicy, sugary, and richly flavoured.

An excellent pear, in use from February till March.

The tree is a hardy and vigorous grower, and forms a handsome pyramid.

Prince's Pear. See *Chair à Dames.*

Princesse de Parme. See *Marie Louise.*

Pucelle Condesienne. See *Passe Colmar.*

Pyrole. See *Jaminette.*

RAMEAU (*Surpasse Reine*).—Fruit large, oblong-oval, and uneven in its outline. Skin lemon-yellow, mottled and dotted with russet. Eye open, slightly depressed. Stalk about an inch long, not depressed. Flesh yellowish, half-melting, juicy, sweet, and perfumed. In use from January till March.

Red Achan. See *Achan.*

RED DOYENNÉ (*Doyenné d'Automne; Doyenné Crotté; Doyenné Galleux; Doyenné Gris; Doyenné Jaune; Doyenné Rouge; Doyenné Roux; Grey Doyenné; Neige Grise; St. Michel Doré; St. Michel Gris*). — Fruit medium sized, obovate. Skin yellowish-green, but entirely covered with thin, smooth, cinnamon-coloured russet, and sometimes with a brownish-red tinge on the side next the sun. Eye small and closed, set in a narrow depression. Stalk three quarters of an inch long, inserted in a narrow, rather deep cavity. Flesh white, tender, melting, very juicy, sugary, and vinous.

A first-rate hardy pear. Ripe in the end of October. The tree is an excellent bearer, and forms a handsome pyramid.

Regentin. See *Passe Colmar.*

REINE DES POIRES.—Fruit medium sized, obovate. Skin smooth, pale yellow, and dotted with russet on the shaded side, and bright red next the sun. Eye small and open, placed in a small, irregular basin. Stalk an inch long, inserted in a small cavity. Flesh yellowish, tender, juicy, and sweet. Ripe in October.

Roberts' Keeping. See *Easter Bergamot.*

Roi Jolimont. See *Doyenné d'Eté.*

Roi de Wurtemburg. See *Glou Morceau.*

RONDELET.—Fruit below medium size, roundish. Skin greenish-yellow, considerably covered with very fine and smooth pale brown russet, having an orange tinge next the sun, and speckled with large grey dots. Eye generally wanting. Stalk an inch long, inserted in a narrow cavity. Flesh fine-grained, buttery and melting, very juicy, rich, sugary, and with a powerful perfume of musk.

A most delicious pear. Ripe in the beginning and middle of November. The tree is quite hardy, an excellent bearer, and succeeds well as a standard or pyramid.

Rose. See *Summer Rose.*

Rosteitzer.—Fruit small or below medium size, pyriform. Skin yellowish-green, with reddish-brown on the exposed side. Eye open, set in a shallow, plaited basin. Stalk an inch and a half long, not depressed. Flesh melting, very juicy, sugary, vinous, and aromatic. Ripe in the end of August and beginning of September.

Rousse Lench.—Fruit large, oblong or oval. Skin pale green, changing to lemon-yellow, with a slight russety covering. Eye large and open, like that of a Jargonelle. Stalk an inch and a quarter long, inserted without depression. Flesh yellow, buttery, juicy, sugary, and pretty well flavoured. Ripe in January and February.

Rousselet d'Anjou. See *Bezi de Caissoy.*

Rousselet Enfant Prodigue.—Fruit medium sized, pyriform. Skin green, considerably covered with rough-brown russet, and with a brownish-red tinge on the exposed side. Eye large and open, set in a shallow basin. Stalk about an inch long, obliquely inserted without depression. Flesh greenish-white, melting, very juicy and sugary, and with a rich, vinous, and musky flavour.

An excellent pear. Ripe in December. The tree is hardy, a good bearer, and forms a handsome pyramid.

Rousselet Jamain. See *Heliote Dundas.*

Rousselet de Meestre.—Fruit large, obtuse-pyriform or pyramidal. Skin smooth and shining, of a golden yellow colour, thickly dotted all over with large brown russet freckles. Eye open, set in a wide, flat basin. Stalk an inch and a half long, not depressed. Flesh half buttery, firm, pretty juicy, and well flavoured, but with nothing to recommend it. Ripe in October and November.

Rousselet Musqué. See *Rousselet de Rheims.*

Rousselet Petit. See *Rousselet de Rheims.*

Rousselet de Rheims (*Rousselet Musqué; Rousselet Petit*).—Fruit small, pyriform, and rounded at the apex. Skin green, changing to yellow at maturity, and thickly covered with grey russet specks, tinged with brown next the sun. Eye small and open, slightly depressed. Stalk an inch long, thick and not depressed. Flesh half-melting, rich, sugary, and highly perfumed.

One of the oldest and best early pears. Ripe in September, but does not keep long.

Rousselet de Stuttgardt.—Fruit medium sized, pyriform or pyramidal. Skin yellowish-green, with brownish-red on the side next the sun, and strewed with dots. Eye open, set in a shallow basin. Stalk upwards of an inch long, inserted without depression. Flesh half-melting, very juicy and sugary, with a rich and perfumed flavour.

A good early pear. Ripe in September. The tree is an excellent bearer, and forms a handsome pyramid.

Royal d'Angleterre. See *Uvedale's St. Germain.*

Royal Tairlon. See *Easter Bergamot.*

Sabine. See *Jaminette.*

Saffran d'Automne. See *Spanish Bon Chrêtien.*

Saffran d'Eté. See *Summer Bon Chrêtien.*

St. Denis.—Fruit small, turbinate, and uneven in its outline. Skin pale yellow, with a crimson cheek, and thickly dotted with crimson dots. Eye open, set in a shallow basin. Stalk an inch and a half long, not depressed. Flesh half-melting, very juicy and sweet, with a fine aroma.

A nice early pear. Ripe in August and September.

St. Germain (*Arteloire; Inconnue la Fare; Lafare; St. Germain Gris; St. Germain d'Hiver; St. Germain Jaune; St. Germain Vert*).—Fruit large, oblong-obovate, rather irregular in its outline. Skin pale greenish-yellow, thickly covered with small brownish-grey dots and sometimes tracings of russet. Eye open, set in a narrow, uneven depression. Stalk an inch long, curved, and inserted without depression. Flesh white, very juicy, buttery and melting, with a sprightly refreshing sugary and perfumed flavour.

A fine old dessert pear, in use from November till January. The tree requires to be grown against a wall.

St. Germain d'Eté. See *Summer St. Germain.*

St. Germain Gris. See *St. Germain.*

St. Germain d'Hiver. See *St. Germain.*

St. Germain Jaune. See *St. Germain.*

St. Germain de Martin. See *Summer St. Germain.*

St. Germain Vert. See *St. Germain.*

St. Ghislain.—Fruit medium size, obtuse-pyriform or turbinate. Skin smooth, clear yellow, with a greenish tinge, and with a blush of red next the sun. Eye open, slightly depressed. Stalk an inch to an inch and a half long, inserted without depression. Flesh white, very juicy, buttery and melting, rich, sugary, and vinous.

An excellent pear. Ripe in September.

St. Jean. See *Amiré Joannet.*

St. Lambert. See *Jargonelle.*

St. Lézin.—Fruit very large, pyriform. Skin of a dull greenish-yellow colour, covered with flakes of russet. Eye open, set in a deep furrowed basin. Stalk two inches long, not depressed. Flesh firm, crisp, juicy, and sweet.

A stewing pear, in use during September and October.

St. Marc. See *Urbaniste.*

St. Martial. See *Angélique de Bordeaux.*

St. Martin. See *Winter Bon Chrêtien.*

St. Michel. See *White Doyenné.*

St. Michel Archange.—Fruit above medium size, obovate. Skin smooth and shining, of a golden-yellow colour, speckled with crimson on the shaded side, and with a bright crimson cheek on the side next the sun. Eye small and closed, set in a narrow depression. Stalk half an inch to an inch long, not depressed. Flesh yellowish-white, tender, melting and juicy, with a sugary juice, and a very agreeable perfume.

A very excellent and beautiful pear, covered with crimson dots like Forelle. Ripe in the end of September.

St. Michel Doré. See *Red Doyenné.*

St. Michel Gris. See *Red Doyenné.*

St. Nicholas. See *Duchesse d'Orleans.*

St. Samson. See *Jargonelle.*

Scotch Bergamot. See *Hampden's Bergamot.*

Scot's Cornuck. See *Charnock.*

SECKLE (*New York Red-cheek; Shakespear; Sicker*). —Fruit small, obovate. Skin yellowish-brown on the shaded side, and reddish-brown next the sun. Eye small and open, not depressed. Stalk half an inch long, inserted in a narrow depression. Flesh buttery, melting, and very juicy, very sweet and rich, with a powerful aroma.

A most delicious pear. Ripe in October. The tree is an abundant bearer, and very hardy; but does not succeed well on the quince.

Seigneur. See *Fondante d'Automne.*

Seigneur. See *White Doyenné.*

Seigneur d'Esperen. See *Fondante d'Automne.*

Seigneur d'Hiver. See *Easter Beurré.*

Sept en Gueule. See *Petit Muscat.*

Serrurieur d'Automne. See *Urbaniste.*

SEUTIN.—Fruit medium sized, oval. Skin yellowish, covered with flakes and dots of russet. Eye prominent and open. Stalk an inch and a half long. Flesh half-melting, coarse-grained, pretty juicy and sweet. Ripe in December and January.

Shakespear. See *Seckle.*

SHOBDEN COURT.—Fruit below medium size, oblate, even in its outline. Skin deep, rich yellow, with a blush of red next the sun, and covered with rough russety dots. Eye very small, almost wanting, set in a small, round, rather deep basin. Stalk very long and slender, inserted in a small cavity. Flesh white, coarse-grained, juicy, briskly acid and sweet, but not highly flavoured. Ripe in January and February.

Short's St. Germain. See *Summer St. Germain.*

Sicker. See *Seckle.*

SIEULLE (*Bergamotte Sieulle; Beurré Sieulle; Doyenné Sieulle*).—Fruit medium sized, roundish-turbinate. Skin smooth, pale yellow, thickly covered with russet dots, and sometimes with a tinge of red next the sun. Eye open, set in a shallow basin. Stalk an inch long, set in a small cavity. Flesh coarse-grained, buttery, and very juicy, rich, sugary, vinous, and aromatic. Ripe in October and November.

SIMON BOUVIER.—Fruit below medium size, obtuse-pyriform. Skin smooth, bright green, becoming yellowish as it ripens, and dotted and mottled with brown russet. Eye small, placed in a slight depression. Stalk three quarters of an inch long, slightly curved, and inserted without depression. Flesh white, tender, and melting, rich, sugary, and finely perfumed. Ripe in September.

Small Winter Beurré. See *Bezi de Caissoy*.

Snow. See *White Doyenné*.

SOLDAT ESPEREN.—Fruit large, obovate. Skin pale lemon-yellow, marked here and there with tracings of russet, and considerably covered with minute dots. Eye large, slightly closed, and placed in a shallow depression. Stalk an inch long, inserted in a narrow cavity. Flesh yellowish-white, buttery, melting, and very juicy, rich and sugary, having somewhat of the flavour of the Autumn Bergamot.

A very excellent pear. Ripe in November.

Souveraine. See *Passe Colmar*.

SPANISH BON CHRÊTIEN (*Gratioli d'Automne; Saffran d'Automne; Spanish Warden*).—Fruit large, pyriform. Skin greenish-yellow, covered with cinnamon-coloured russet, and with a deep lively red colour next the sun. Eye open, set in a depression. Stalk an inch and a half long, slender, inserted without depression. Flesh white, fine-grained and crisp, with a brisk flavour, and fine musky aroma.

A fine stewing pear, in use from November till March.

Spanish Warden. See *Spanish Bon Chrêtien*.

DE SPOELBERG.—Fruit about medium size, somewhat turbinate and uneven in its outline, being considerably ribbed and undulating. Skin smooth, pale straw coloured, sprinkled with green dots and patches of russet. Eye large, half open, and prominently set. Stalk an inch and a quarter long, inserted without depression. Flesh yellowish, buttery, not very juicy, sweet, slightly musky, and richly flavoured. Ripe in November.

Spring Beurré. See *Verulam*.

Staunton. See *Gansel's Bergamot*.

SUCRÉE VERT (*Green Sugar*).—Fruit medium sized, roundish-turbinate. Skin pale yellowish-green, covered with numerous green and grey dots, and a few tracings of russet. Eye small and open, set in a wide and shallow basin. Stalk an inch long, inserted in a small cavity. Flesh yellowish-white, melting and very juicy, sugary and perfumed. Ripe in October.

SUFFOLK THORN.—Fruit medium sized, roundish-turbinate. Skin pale lemon-yellow, covered with numerous small dots and irregular patches of pale ashy-grey russet, which are most numerous on the side next the sun. Eye very small and open, set in a deep basin. Stalk short and stout, not deeply inserted. Flesh yellowish-white, exceedingly melting, buttery, and juicy, with a rich sugary juice exactly similar in flavour to Gansel's Bergamot.

A most delicious pear. Ripe in October. The tree is quite hardy, and an excellent bearer, forming a handsome pyramid on the pear stock.

SUMMER ARCHDUKE (*Amiré Roux; Archduke d'Eté; Brown Admiral; Grosse Ognonet; Ognonet; Ognonet Musqué*).—Fruit medium sized, turbinate. Skin smooth and shining, yellowish-green, covered with dark brownish-red next the sun. Eye open, set in a shallow depression. Stalk an inch long, stout, inserted in a small cavity. Flesh whitish, rather gritty, juicy, and sweet. Ripe in the beginning of August.

The Summer Rose is also called *Ognonet*.

Summer Bell. See *Windsor*.

SUMMER BON CHRÊTIEN (*Florence d'Eté; Gratioli; Gratioli di Roma; Saffran d'Eté*).—Fruit large, pyriform, very irregular and bossed in its outline. Skin yellow, with a tinge of pale red next the sun, and strewed with green specks. Eye small, set in an uneven, shallow basin. Stalk two inches and a half long, curved, and obliquely inserted in a knobbed cavity. Flesh yellow, crisp, juicy, sweet, and pleasantly flavoured. Ripe in September.

SUMMER CRASANNE (*Crasanne d'Eté*).—Fruit small, roundish, and flattened. Skin pale yellow, entirely covered with cinnamon-coloured russet. Eye wide open, set in a shallow basin. Stalk an inch and a half long. Flesh half-melting, very juicy, sweet, and aromatic. Ripe in the end of August and beginning of September.

SUMMER DOYENNÉ (*Doyenné d'Eté; Doyenné de Juillet; Duchesse de Berri d'Eté; Roi Jolimont*).—Fruit small, roundish-obovate. Skin smooth, of a fine yellow colour, and frequently with a red blush on the side next the sun, and strewed with dots. Eye small and open, set in a shallow plaited basin. Stalk short, not depressed. Flesh white, melting, and very juicy, rich and sugary.

An excellent early pear. Ripe in the end of July, but requires to be gathered before it becomes yellow, otherwise it soon decays. The tree is hardy, and a good bearer.

SUMMER FRANC REAL (*Coule Soif; Fin Or d'Eté; Franc Real d'Eté; Great Mouthwater; Gros Micet d'Eté; Grosse Mouille Bouche*).—Fruit medium sized, obovate. Skin smooth, pale yellowish-green, strewed with numerous brown and green dots. Eye small and open, set in a small undulating basin. Stalk short and thick, inserted in a small cavity. Flesh white, fine-grained, buttery and melting, rich and sugary.

An excellent early pear. Ripe in September.

SUMMER ROSE (*Epine Rose; Ognonet; Rose; Thorny Rose*).—Fruit medium sized, oblate. Skin greenish-yellow on the shaded side, and bright reddish-purple on the side next the sun, strewed with russet dots. Eye open, set in a wide and shallow basin. Stalk an inch and a half long, slender, inserted in a small cavity. Flesh half-melting, tender, juicy, sugary, with a pleasant, refreshing flavour and musky aroma.

A very nice early pear. Ripe in August.

SUMMER ST. GERMAIN (*St. Germain d'Eté; St. Germain de Martin; Short's St. Germain*).—Fruit medium sized, obovate. Skin greenish pale yellow, mottled and speckled with brown russet. Eye open, set in a narrow and slight depression. Stalk an inch and a quarter long, inserted in a small cavity. Flesh juicy, slightly gritty and astringent, with a brisk, sweet, and rather pleasant flavour.

A second-rate pear. Ripe in the end of August.

SUMMER THORN (*Bugiarda; Epine d'Eté Couleur de Rose; Epine d'Eté Vert; Fondante Musqué*).—Fruit medium sized, pyriform or long pyriform, and rounded at the apex. Skin smooth, and covered with greenish-russet dots, green in the shade, but yellowish next the sun and towards the stalk. Eye small, set in a shallow and plaited basin. Stalk an inch long, curved, and obliquely inserted without any depression. Flesh white, melting, juicy, and of a rich musky flavour.

It is an excellent autumn pear, ripe in September, but does not keep long.

Suprême. See *Windsor.*

Surpasse Reine. See *Rameau.*

Suzette de Bavay.—Fruit medium sized, turbinate. Skin yellow, covered with numerous large russet dots and traces of russet. Eye open, placed in a shallow, undulating basin. Stalk an inch long, inserted in a small cavity. Flesh melting, juicy, sugary, and vinous, with a pleasant perfume. Ripe in January and February.

Swan's Egg.—Fruit medium sized, roundish-ovate. Skin smooth, yellowish-green on the shaded side, and clear brownish-red next the sun, and covered with pale brown russet. Eye small, partially closed, slightly depressed. Stalk an inch and a half long, inserted without depression. Flesh tender, very juicy, with a sweet and piquant flavour and musky aroma.

A fine old variety. Ripe in October. The tree is very hardy, and an excellent bearer.

Sweet Summer. See *Jargonelle.*

Sylvange d'Hiver. See *Easter Beurré.*

Table des Princes. See *Jargonelle.*

Tardif de Mons.—Fruit oblong-obovate, even and regularly formed. Skin of a uniform yellow colour, paler on the shaded side, and with an orange tinge next the sun, strewed with large russety dots. Eye open, very slightly depressed. Stalk an inch long, rather slender, not depressed. Flesh white, tender, buttery, melting, and very juicy, rich and sugary Ripe in November.

Tarling. See *Easter Bergamot.*

Têton de Vénus. See *Bellissime d'Hiver.*

Têton de Vénus. See *Catillac.*

Théodore Van Mons.—Fruit large, pyramidal. Skin greenish-yellow, strewed with russety dots and tracings of russet. Eye closed, set in a small, uneven basin. Stalk three quarters of an inch long, inserted without depression. Flesh yellowish-white, juicy and melting. Ripe in October and November.

Thessoise. See *Beurré d'Amanlis.*

THOMPSON'S.—Fruit medium sized, obovate. Skin pale yellow, and considerably covered with a coating and dots of pale cinnamon-coloured russet. Eye open, set in a shallow basin. Stalk an inch and a quarter long, inserted in an uneven cavity. Flesh white, buttery and melting, very juicy, exceedingly rich and sugary, and with a fine aroma.

One of our best pears. Ripe in November. The tree is quite hardy, an excellent bearer, and succeeds best on the pear stock.

TILLINGTON.—Fruit about medium size, short pyriform, rather uneven in its outline. Skin smooth, greenish-yellow, covered with a number of light brown russet dots. Eye open, scarcely at all depressed. Stalk short, fleshy, and warted at its insertion. Flesh yellowish, tender, buttery and melting, not very juicy, but brisk and vinous, with a peculiar and fine aroma.

This is an excellent pear, ripe in October, the fine piquant flavour of which contrasts favourably with the luscious sweetness of the Seckle, which comes in just before it.

Tombe de l'Amateur. See *Nouveau Poiteau.*

De Tonneau. See *Uvedale's St. Germain.*

Tres Grosse de Bruxelles. See *Uvedale's St. Germain.*

Triomphe de Hasselt. See *Calebasse Grosse.*

TRIOMPHE DE JODOIGNE.—Fruit large, obovate, regular and handsome. Skin yellow, covered with numerous small russety dots and patches of thin brown russet. Eye open, set in a slight depression. Stalk an inch and a quarter long, curved, and inserted without depression. Flesh yellowish-white, rather coarse, melting, juicy. sugary, and brisk, with an agreeable musky perfume. Ripe in November and December.

TRIOMPHE DE LOUVAIN.—Fruit medium sized, obovate. Skin covered with fawn-coloured russet, and densely strewed with light-brown russet dots; except on the exposed side, where it is of a deep dull red. Eye open, set in a shallow basin. Stalk an inch long, thick, with a fleshy protuberance on one side. Flesh white, crisp, juicy, and sweet; but decays at the core before it begins to melt. Ripe in the end of September.

De Trois Tours. See *Beurré Diel.*

Trompe Valet. See *Ambrette d'Hiver.*

Trout. See *Forelle.*

Truite. See *Forelle.*

Union. See *Uvedale's St. Germain.*

URBANISTE (*Beurré Drapiez; Beurré Picquery; Louise d'Orleans; Picquery; St. Marc; Serrurier d'Automne; Virgalieu Musquée*).—Fruit medium sized, obovate, or oblong-obovate. Skin smooth and thin, pale yellow, covered with grey dots and slight markings of russet, and mottled with reddish brown. Eye small and closed, set in a deep, narrow basin. Stalk an inch long, inserted in a wide and rather deep cavity. Flesh white, very tender, melting, and juicy, rich, sugary, and slightly perfumed.

A delicious pear. Ripe in October. The tree is hardy and an excellent bearer, forming a handsome pyramid either on the pear or the quince.

UVEDALE'S ST. GERMAIN (*Abbé Mongein; Angora; Belle de Jersey; Bolivar; Chambers' Large; Comtesse de Treweren; Dr. Udale's Warden; Duchesse de Berri d'Hiver; German Baker; Lent St. Germain; Pickering Pear; Pickering's Warden; Piper; Royale d'Angleterre; De Tonneau; Tres Grosse de Bruxelles; Union*).—Fruit very large, sometimes weighing upwards of 3 lbs., of a long pyriform or pyramidal shape. Skin smooth, dark green, changing to yellowish-green, and with dull brownish-red on the exposed side, dotted all over with bright brown and a few tracings of russet. Eye open, set in a deep, narrow cavity. Stalk an inch to an inch and a half long, inserted in a small cavity. Flesh white, crisp, and juicy.

An excellent stewing pear, in use from January to April.

VALLÉE FRANCHE (*Bonne de Kienzheim; De Kienzheim*).—Fruit medium sized, obovate or obtuse-pyriform. Skin smooth and shining, yellowish-green, becoming yellowish as it ripens, and covered with numerous small russet dots. Eye set in a shallow basin. Stalk an inch long, inserted without depression. Flesh white, rather crisp, very juicy and sweet.

A good early pear. Ripe in the end of August. The tree is an immense and regular bearer, very hardy, and an excellent orcharding variety.

Van Assche.—Fruit large, roundish-oval, bossed and ribbed in its outline. Skin yellow, covered with flakes of russet on the shaded side, and with beautiful red on the side next the sun. Eye half open, set in a ribbed basin. Stalk half an inch long, inserted in a small cavity. Flesh half-melting, very juicy, rich, and aromatic. In use during November and December.

Van Donckelaar. See *Marie Louise.*

Van Marum. See *Grosse Calebasse.*

Van Mons Léon le Clerc.—Fruit very large, oblong-pyramidal. Skin dull yellow, covered with dots and tracings of russet. Eye open, set in a shallow basin. talk an inch and a half long, curved, and inserted in a shallow cavity. Flesh yellowish-white, buttery and melting, very juicy, rich, sugary, and delicious.

A remarkably fine pear. Ripe in November. The tree is an excellent bearer, succeeds well as a standard in warm situations, and forms a handsome pyramid on the pear stock.

Van de Weyer Bates.—Fruit below medium size, roundish-obovate. Skin pale lemon yellow, covered with small brown dots and a few veins of russet of the same colour. Eye very large and open, set in a moderate depression. Stalk an inch and a quarter long, inserted between two lips. Flesh yellow, buttery, and very juicy, rich and sugary, with a pleasant aroma.

One of the finest late pears. Ripe from March till May

Vergalieu Musquée. See *Urbaniste.*

Verlaine. See *Gendesheim.*

Verlaine d'Eté. See *Gendesheim.*

Vert Longue (*Mouille Bouche; Mouille Bouche d'Automne; New Autumn*).—Fruit medium sized, pyriform. Skin smooth and shining, pale green, becoming yellowish about the stalk as it ripens, and covered with numerous minute dots. Eye open, set in a shallow basin. Stalk an inch and a half long, not depressed. Flesh white, melting, very juicy, sugary, and richly flavoured. Ripe in October.

Verte Longue d'Hiver. See *Echassery.*

Verulam (*Black Beurré; Buchanan's Spring Beurré; Spring Beurré*).—Fruit large, obovate, resembling the Brown Beurré in shape. Skin dull green, entirely covered with thin russet on the shaded side, and reddish-brown thickly covered with grey dots on the side next the sun. Eye open, set in a shallow basin. Stalk an inch long, slender, inserted in a small cavity. Flesh crisp, coarse-grained, rarely melting, unless grown against a wall in a warm situation, which is a position it does not merit.

An excellent stewing pear, in use from January till March. When stewed the flesh assumes a fine brilliant colour, and is richly flavoured.

Viandry. See *Echassery*.

VICAR OF WINKFIELD (*Belle Andrenne; Belle de Berri; Belle Heloise; Beurré Comice de Toulon; Bon Papa; Curé; Monsieur de Clion; Monsieur le Curé; Paternoster*).—Fruit very large, pyriform, frequently one-sided. Skin smooth, greenish-yellow, with a faint tinge of red on the side next the sun. Eye open, set in a shallow basin, and placed on the opposite side of the axis from the stalk. Stalk an inch and a half long, slender, obliquely inserted without depression. Flesh white, fine-grained, half-melting, juicy and sweet, with a musky aroma.

A handsome pear, which in warm seasons, or when grown against a wall, is melting. It is also a pretty good stewing pear. In use from November till January.

Vigne de Pelone. See *Figue de Naples*.

VINEUSE.—Fruit medium sized, obovate. Skin smooth, pale straw colour, with slight markings of very thin brown russet, interspersed with minute green dots. Eye open, frequently abortive, set in a shallow depression. Stalk short and fleshy, inserted in a deep, narrow cavity. Flesh yellowish-white, exceedingly tender, melting, and very juicy, of a honied sweetness, and fine delicate perfume.

A delicious and richly-flavoured pear. Ripe in the end of September and beginning of October.

VINGOULEUSE (*Bujaleuf; Chambrette; De Glace*).—Fruit large and pyriform. Skin smooth and delicate, pale lemon colour, with a tinge of brown on the side next the sun, thickly strewed with russet dots. Eye small and open, set in a small, narrow basin. Stalk an inch to an inch and a quarter long, inserted without depression. Flesh yellowish-white, buttery, melting, and very juicy, sugary, and perfumed. November till January.

Warwick Bergamot. See *White Doyenné*.

WELBECK BERGAMOT.—Fruit above medium size, roundish, uneven in its outline, and bossed about the stalk. Skin smooth and shining, of a lemon-yellow colour, thickly sprinkled with large russet specks and with a blush of light crimson on the side next the sun. Eye small and open, set in a shallow depression. Stalk three quarters of an inch long, inserted in an uneven cavity. Flesh white, rather coarse-grained, half-melting, very juicy, sweet, and sugary, but without any flavour. End of October and November.

White Autumn Beurré. See *White Doyenné.*

White Beurré. See *White Doyenné.*

WHITE DOYENNÉ (*Beurré Blanc; Bonne Ente; Citron de Septembre; Dean's; Doyenné Blanc; Doyenné Picté; Neige; Pine; St. Michel; Seigneur; Snow; Warwick Bergamot; White Autumn Beurré; White Beurré*).—Fruit above medium size, obovate, handsome, and regularly formed. Skin smooth and shining, pale straw colour, sometimes with a faint tinge of red next the sun, and strewed with small dots. Eye very small and closed, set in a small, shallow basin. Stalk three quarters of an inch long, stout, fleshy, set in a small, round cavity. Flesh white, fine-grained, buttery, and melting, rich, sugary, with a fine piquant and vinous flavour, and a delicate perfume.

A delicious fruit. Ripe in September and October. The tree is hardy, a free bearer, and succeeds well as a pyramid either on the pear or quince.

Wilding of Caissoy. See *Bezi de Caissoy*

Wilhelmine. See *Beurré d'Amanlis.*

WILLERMOZ.—Fruit large, obtuse-pyriform, ribbed and bossed in its outline. Skin of a golden yellow colour, with a red blush on the exposed side, and covered with fine russet dots. Stalk an inch long, woody. Flesh white, fine-grained, buttery, and melting, very juicy, sugary, and highly perfumed. Ripe in October and November.

William the Fourth. See *Louise Bonne of Jersey.*

Williams'. See *Williams' Bon Chrêtien.*

WILLIAMS' BON CHRÊTIEN (*Bartlett; De Lavault; Williams'*).—Fruit large, obtuse-pyriform, irregular and bossed in its outline. Skin smooth, of a fine clear yellow, tinged with green mottles and with faint streaks of red on the exposed side. Eye open, set in a shallow depression. Stalk an inch long, stout and fleshy, inserted in a shallow cavity, which is frequently swollen on one side. Flesh white, fine-grained, tender, buttery, and melting, with a rich, sugary, and delicious flavour, and powerful musky aroma.

One of the finest of pears. Ripe in August and September. It should be gathered before it becomes yellow, otherwise it speedily decays. The tree forms a handsome pyramid, and is a good bearer.

WINDSOR (*Bell Tongue; Bellissime; Figue; Figue Musquée; Green Windsor; Grosse Jargonelle; Konge; Madame; Madame de France; Summer Bell; Suprême*). —Fruit large, pyriform, rounded at the eye. Skin smooth, green at first, and changing to yellow mixed with green, and with a faint tinge of orange and obscure streaks of red on the exposed side. Eye open, not at all depressed. Stalk an inch and a half long, inserted without depression. Flesh white, tender, buttery, and melting, with a fine, brisk, vinous flavour, and nice perfume.

A fine old pear for orchard culture. Ripe in August. It should be gathered before it becomes yellow.

Winter Beurré. See *Achan*.

Winter Beurré. See *Chaumontel*.

WINTER BON CHRÊTIEN (*D'Angoisse; Bon Chrêtien d'Hiver; Bon Chrétien de Tours; De St. Martin*).—Fruit large, obtuse-pyriform, very irregular and bossed in its outline. Skin dingy yellow, with a tinge of brown next the sun, and strewed with small russet dots. Eye open, set in a deep basin. Stalk an inch to an inch and a half long, inserted in a small cavity. Flesh white, crisp, juicy, sweet, and perfumed.

This requires a wall, but is not worthy of such a situation. It is in use from December to March; and is more adapted for stewing than for the dessert.

WINTER FRANC REAL (*Fin Or d'Hiver; Franc Real d'Hiver; Gros Micet*).—Fruit medium sized, obovate, uneven in its outline. Skin of a fine lemon-yellow colour, with light brownish-red next the sun, thickly covered with pale brown dots and markings of russet. Eye open, set in a rather deep basin. Stalk an inch long, inserted in a deep cavity. Flesh yellowish, coarse-grained, juicy, sweet, and aromatic.

A fine stewing pear, in use from January till March. When cooked the flesh becomes of a fine bright purple colour, and richly flavoured.

Winter Green. See *Bergamotte d'Hollande.*

WINTER OKEN (*Oken; Oken d'Hiver*).—Fruit below medium size, roundish. Skin lemon yellow, marked with patches of cinnamon-coloured russet. Eye open, set in a round, deep basin. Stalk an inch long, inserted without depression. Flesh buttery, melting, and juicy, rich, sugary, and well flavoured. Ripe in December.

WINTER NELIS (*Beurré de Malines; Bonne Malinaise; Bonne de Malines; Colmar Nelis; Etonneau; Fondante de Malines; Malinaise Cuvelier; Nelis d'Hiver*).—Fruit below medium size, roundish-obovate. Skin dull yellowish-green, covered with numerous russety dots and patches of brown russet. Eye open, set in a shallow depression. Stalk from an inch to an inch and a half long, set in a narrow cavity. Flesh yellowish, fine-grained, buttery and melting, with a rich, sugary, and vinous flavour, and a fine aroma.

One of the richest flavoured pears. It is in use from November till February. The tree forms a handsome small pyramid, is quite hardy, and an excellent bearer.

Winter Poplin. See *Bezi de Caissoy.*

WINTER THORN (*Epine d'Hiver; Epine Rose d'Hiver*).—Fruit above medium size, obovate. Skin smooth, yellowish-green, covered with greyish-brown dots. Eye small and open, set in a wide basin. Stalk an inch long, inserted without depression. Flesh whitish, tender, and buttery, with a sweet and agreeable musky flavour. In use from November till January.

WINTER WINDSOR (*Petworth*). — Fruit large and handsome, obovate-turbinate. Skin smooth and shining, greenish yellow in the shade and orange, faintly streaked with brownish-red next the sun; covered all over with minute dots. Eye large and open, set in a shallow basin. Stalk half an inch long, slender, inserted without depression. Flesh crisp, juicy, and pleasantly flavoured. Ripe in November.

YAT (*Yutte*).—Fruit below medium size, obtuse-pyriform. Skin thickly covered with brown russet and sprinkled with numerous grey specks, sometimes with brownish-red next the sun. Eye small and open, set in a shallow basin. Stalk an inch long, obliquely inserted without depression. Flesh white, tender, juicy, and melting, with a rich, sugary, and highly perfumed flavour.

An excellent early pear. Ripe in September. The tree is hardy, and a great bearer.

York Bergamot. See *Autumn Bergamot.*

Yutte. See *Yat.*

ZÉPHIRIN GRÉGOIRE. — Fruit about medium size, oundish. Skin pale greenish-yellow, sometimes becoming of a uniform pale waxen yellow, covered with russet dots and markings. Eye very small, slightly depressed. Stalk an inch long, inserted without depression. Flesh yellow, buttery, melting, and very juicy, very rich, sugary, and vinous, with a powerful and peculiar aroma.

A most delicious pear. Ripe in December and January. The tree forms a handsome pyramid, succeeds best on the pear stock, and is an excellent bearer.

---

## LISTS OF SELECT PEARS,

*Arranged in their order of ripening.*

### I. COLLECTIONS OF SIX VARIETIES FOR PYRAMIDS, BUSHES, OR ESPALIERS.

1.

Jargonelle
Williams' Bon Chrêtien
Urbaniste
Soldat Esperen
Catinka
Ne Plus Meuris

2.

Citron des Carmes
Louise Bonne of Jersey
Jersey Gratioli
Nouveau Poiteau
Rousselet Enfant Prodigue
Beurré Sterckmans

3.

Beurré Giffard
Beurré d'Amanlis
Baronne de Mello
Van Mons Léon le Clerc
Doyenné Defais
Glou Morceau

4.

Bloodgood
Beurré Superfin
Seckle
Marie Louise
Knight's Monarch
Beurré de Rance

5.

Hampden's Bergamot
Fondante d'Automne
Paradise d'Automne
Rondelet
Winter Nelis
Joséphine de Malines

6.

Summer Rose
Navez Peintre
Duchesse d'Orléans
Figue de Naples
Jewess
Zéphirin Grégoire

7.

Ambrosia
Albertine
Comte de Lamy
Beurré Berckmans
Moccas
L'Inconnue

8.

Flemish Beauty
Henry the Fourth
Eyewood
Thompson's
Beurré Duval
Forelle

9.

Early Rousselet
Beurré Goubault
Red Doyenné
Suffolk Thorn
Henriette Bouvier
Huyshe's Bergamot

## II. COLLECTIONS OF TWELVE VARIETIES FOR PYRAMIDS, BUSHES, OR, ESPALIERS.

1.

Citron des Carmes
Hampden's Bergamot
Beurré d'Amanlis
Louise Bonne of Jersey
Seckle
Van Mons Léon le Clerc
De Spoelberg
Dr. Trousseau
Beurré Berckmans
Winter Nelis
Beurré Sterckmans
Easter Beurré.

2.

Doyenné d Eté
Jargonelle
Beurré Giffard
Williams' Bon Chrêtien
Albertine
Beurré Hardy
Beurré Diel
Soldat Esperen
Henriette Bouvier
Glou Morceau
Ne Plus Meuris
Bergamotte Esperen

3.

| | |
|---|---|
| Early Rousselet | Duchesse d'Angoulême |
| Summer Rose | Nouveau Poiteau |
| Flemish Beauty | Beurré Bosc |
| Peach | Jewess |
| Henry the Fourth | Moccas |
| Baronne de Mello | Zéphirin Grégoire |

4.

| | |
|---|---|
| Bloodgood | Suffolk Thorn |
| St. Denis | Thompson's |
| Beurré Superfin | Catinka |
| Fondante d'Automne | Knight's Monarch |
| Comte de Lamy | Joséphine de Malines |
| Marie Louise | Van de Weyer Bates |

5.

| | |
|---|---|
| Yat | Beurré Duval |
| Beurré Goubault | Doyenné Defais |
| Vineuse | Forelle |
| Jersey Gratioli | Huyshe's Bergamot |
| Eyewood | L'Inconnue |
| Rondelet | Beurré de Rance |

6.

| | |
|---|---|
| Vallée Franche | Figue de Naples |
| Beurré Benoît | Comte de Flandres |
| Navez Peintre | Rousselet Enfant Prodigue |
| Doyenné Boussoch | Alexandre Bivort |
| Duchesse d'Orléans | Jean de Witte |
| Paradise d'Automne | Cassante de Mars |

III. VARIETIES REQUIRING A WALL, OR WHICH ARE IMPROVED BY SUCH PROTECTION.

| | |
|---|---|
| Bergamotte Esperen | Forelle |
| Beurré Bosc | Gansel's Bergamot |
| Beurré Diel | Glou Morceau |
| Beurré de Rance | Knight's Monarch |
| Beurré Sterckmans | Ne Plus Meuris |
| Brown Beurré | Passe Colmar |
| Colmar | Prince Albert |
| Crassanne | St. Germain |
| Duchesse d'Angoulême | Van Mons Léon le Clerc |
| Easter Beurré | Winter Nelis |

### IV. VARIETIES FOR ORCHARD STANDARDS.

Aston Town
Autumn Bergamot
Beurré Capiaumont
Bishop's Thumb
Caillot Rosat
Croft Castle
Eyewood
Hampden's Bergamot
Hessle
Jargonelle
Jersey Gratioli
Lammas
Louise Bonne of Jersey
Suffolk Thorn
Swan's Egg
Williams' Bon Chrêtien
Windsor
Winter Nelis
Vallée Franche
Yat

### V. VARIETIES FOR STEWING AND PRESERVING.

Belmont
Bezi d'Heri
Black Worcester
Catillac
Flemish Bon Chrêtien
Gilogil
Verulam
Winter Franc Real

### VI. VARIETIES FOR NORTHERN LATITUDES, AND EXPOSED SITUATIONS IN THE MIDLAND AND SOUTHERN COUNTIES.

*Those marked * require a wall.*

Doyenné d'Eté
Citron des Carmes
Jargonelle
Williams' Bon Chrêtien
Beurré d'Amanlis
Louise Bonne of Jersey
Hessle
Comte de Lamy
Jersey Gratioli
Red Doyenné
Thompson's
Duchesse d'Angoulême*
Marie Louise*
Beurré Diel*
Knight's Monarch
Beurré de Rance*

# PLUMS.

## SYNOPSIS OF PLUMS.

### I. FRUIT ROUND.

** Summer shoots smooth.*

1. SKIN DARK.

A. *Freestones.*

Angelina Burdett
Corse's Nota Bene
Damas de Mangeron
Italian Damask
Kirke's
Late Orleans
De Montfort
Nectarine
Peach
Purple Gage
Queen Mother
Woolston Black Gage

B. *Clingstones.*

Belgian Purple
Frost Gage
Lombard
Nelson's Victory
Prince of Wales
Suisse

2. SKIN PALE.

A. *Freestones.*

Abricotée de Braunau
Aunt Ann
General Hand
Green Gage
July Green Gage
Late Green Gage
Reine Claude de Bavay
White Damask
Yellow Gage

B. *Clingstones.*

Large Green Drying
Lucombe's Nonesuch
McLaughlin

*** Summer shoots downy.*

1. SKIN DARK.

A. *Freestones.*

Blue Gage
Coe's Late Red
Columbia
Damas Musqué
Damas de Provence
Early Orleans
Orleans
Royale
Royale Hâtive
Royale de Tours
Tardive de Chalons

B. *Clingstones.*

Morocco

2. SKIN PALE.

A. *Freestones.*

Apricot
Denniston's Superb
Drap d'Or
Lawrence Gage

B. *Clingstones.*

Hulings' Superb
Imperial Ottoman
White Bullace

## II. FRUIT OVAL.

### † *Summer shoots smooth.*

#### 1. Skin dark.

A. *Freestones.*

D'Agen
Autumn Compôte
Cooper's Large
Early Prolific
Fotheringham
Italian Quetsche
Quetsche
Red Magnum Bonum
Royal Dauphin
Standard of England

B. *Clingstones.*

Blue Impératrice.
Cherry
Ickworth Impératrice
Impériale de Milan
Pond's Seedling
Prince Englebert
Smith's Orleans

#### 2. Skin pale.

A. *Freestones.*

Autumn Gage
Damas Dronet
Dunmore
Jefferson
Mamelonné
St. Etienne
St. Martin's Quetsche
Transparent Gage
White Impératrice

B. *Clingstones.*

Coe's Golden Drop
Downton Impératrice
Emerald Drop
Guthrie's Apricot
Guthrie's Late Green
Mirabelle Tardive
St. Catherine
Topaz
White Magnum Bonum
Yellow Impératrice

### †† *Summer shoots downy.*

#### 1. Skin dark.

A. *Freestones.*

Cheston
Damas de Septembre
Damson
Diamond
Diaprée Rouge
Early Favourite
Isabella
Perdrigon Violet Hâtif
Red Perdrigon
Reine Claude Rouge
Stoneless
Victoria
Violet Damask

B. *Clingstones.*

Belle de Septembre
Blue Perdrigon
Corse's Admiral
Goliath
Isabella
Précoce de Tours
Prune Damson
Winesour

#### 2. Skin pale.

A. *Freestones.*

Bleeker's Gage
Gisborne's
Imperial Gage
Mirabelle Petite
Précoce de Bergthold
Washington
White Perdrigon
White Primordian

B. *Clingstones.*

White Damson

Abricotée Blanche. See *Apricot.*

ABRICOTÉE DE BRAUNAU.—Fruit about medium size, roundish, and marked with a deep suture. Skin green, like the Green Gage, covered with a white bloom, and becoming yellowish as it ripens, and sometimes with a blush of red next the sun. Stalk an inch long, stout. Flesh greenish-yellow, rather firm in texture, juicy and rich, with a fine and remarkable piquancy, and separating freely from the stone. The kernel is rather sweet. Shoots smooth.

A most excellent plum. Ripe in the beginning of September. Its fine sprightly flavour is as remarkable among dessert plums as that of the Mayduke is among cherries.

Abricotée de Tours. See *Apricot.*

D'AGEN (*Agen Date; Prune D'Ast; Prune du Roi; Robe de Sargent; St. Maurin*).—Fruit medium sized, obovate, and somewhat flattened on one side. Skin deep purple, almost approaching to black, and covered with blue bloom. Stalk short. Flesh greenish-yellow, sweet and well flavoured. Shoots smooth.

An excellent drying and preserving plum. Ripe in September. It is this which, in a dried state, forms the celebrated *pruneaux d'Agen.*

Agen Date. See *D'Agen.*

Alderton. See *Sharp's Emperor.*

Amber Primordian. See *White Primordian.*

American Damson. See *Frost Gage.*

ANGELINA BURDETT.—Fruit above medium size, round, and marked with a suture, which is deepest towards the stalk. Skin thick, dark purple, thickly covered with brown dots and blue bloom. Stalk about an inch long. Flesh yellowish, juicy, rich, and highly flavoured, separating from the stone. Shoots smooth.

An excellent dessert plum. Ripe in the beginning of September, and if allowed to hang till it shrivels, it forms a perfect sweetmeat. The tree is a good bearer and hardy.

Anglaise Noire. See *Orleans.*

APRICOT (*Abricotée Blanche; Abricotée de Tours; Old Apricot; Yellow Apricot*).—Fruit larger than Green

Gage, roundish, and slightly elongated, with a deep suture on one side of it. Skin yellowish, with a tinge of red on the side next the sun, strewed with red dots, and covered with a white bloom. Stalk about half an inch long. Flesh yellow, melting and juicy, with a rich, pleasant flavour, and separating from the stone. Young shoots covered with a whitish down.

A dessert plum, requiring a wall to have it in perfection, and when well ripened little inferior to Green Gage. Ripe in the middle of September.

Askew's Golden Egg. See *White Magnum Bonum.*

Askew's Purple Egg. See *Red Magnum Bonum.*

Aunt Ann (*Guthrie's Aunt Ann*).—This is a large, round plum, of a greenish-yellow colour. The flesh of a rich, juicy flavour, and separates freely from the stone. Shoots smooth.

It ripens in the middle of September. The tree is very hardy and productive.

Autumn Compôte.—This is a very large oval-shaped plum, raised by Mr. Rivers, of Sawbridgeworth, from Cooper's Large. It is very handsome, and the skin is of a bright red colour. As a culinary plum, or for preserving, it is of the first quality. When preserved the pulp is of an amber colour, flavour rich, and possessing more acidity than the Green Gage does when preserved. It is ripe in the end of September. Shoots smooth.

Autumn Gage (*Roe's Autumn Gage*).—Fruit medium sized, oval or rather cordate, marked with a shallow suture, which extends to half the length of the fruit. Skin pale yellow, covered with thin whitish bloom. Stalk three quarters of an inch long, not depressed. Flesh greenish-yellow, juicy and sweet, with a rich and excellent flavour. Shoots smooth.

An excellent dessert plum. Ripe in the middle of October. The tree is an excellent bearer.

Avant Prune Blanche. See *White Primordian.*

D'Avoine. See *White Primordian.*

Azure Hâtive. See *Blue Gage.*

Battle Monument. See *Blue Perdrigon.*

Becker's Scarlet. See *Lombard.*

Beekman's Scarlet. See *Lombard.*

BELGIAN PURPLE (*Bleu de Belgique*).—Fruit medium sized, roundish, marked with a shallow suture, one side of which is a little swollen. Skin deep purple, covered with blue bloom. Stalk about an inch long, inserted in a cavity. Flesh greenish, rather coarse, very juicy, sweet, and rich, slightly adherent to the stone. Shoots smooth. Ripe in the middle of August.

BELLE DE SEPTEMBRE (*Reina Nova; Gros Rouge de Septembre*).—Fruit large, roundish-oval, marked with a shallow suture. Skin thin, violet-red, thickly covered with yellow dots, and a thin blue bloom. Stalk half an inch long, slender, inserted in a shallow cavity. Flesh yellowish-white, firm, juicy, sweet, and aromatic. Shoots downy.

A first-rate plum for cooking or preserving; it furnishes a fine crimson juice or syrup. Ripe in the beginning and middle of October.

Black Damask. See *Morocco*.

Black Morocco. See *Morocco*.

Black Perdrigon. See *Blue Gage*.

Bleeker's Gage. See *Bleeker's Yellow Gage*.

Bleeker's Scarlet. See *Lombard*.

BLEEKER'S YELLOW GAGE (*Bleeker's Gage; German Gage*).—Fruit medium sized, roundish-oval, marked with a faint suture. Skin yellow, containing numerous imbedded white specks, and covered with thin white bloom. Stalk downy, an inch and a quarter long, not depressed. Flesh yellow, rich, and sweet, separating freely from the stone. Shoots downy. Ripe in the middle of September.

Bleu de Belgique. See *Belgian Purple*.

BLUE GAGE (*Azure Hâtive; Black Perdrigon; Cooper's Blue Gage*).—Fruit of medium size, quite round. Skin dark purple, covered with a blue bloom. Stalk three quarters of an inch long. Flesh yellowish-green, juicy, briskly and somewhat richly flavoured, separating from the stone. Shoots downy.

A second-rate plum. Ripe in the beginning of August.

BLUE IMPÉRATRICE (*Impératrice; Empress*).—Fruit medium sized, obovate, tapering considerably towards the stalk, and marked with a shallow suture. Skin deep purple, covered with a thick blue bloom. Stalk about an inch long, not depressed. Flesh greenish-yellow, of a rich sugary flavour, and adhering to the stone. Shoots smooth.

A first-rate plum either for the dessert or preserving. Ripe in October. The tree requires a wall, and the fruit will hang long on the tree, when it becomes shrivelled and very rich in flavour.

Blue Perdrigon (*Brignole Violette; Battle Monument; Perdrigon Violette; Violet Perdrigon*).—Fruit medium sized, oval, widest at the apex, flattened on the side marked with the suture, which is shallow. Skin reddish-purple, marked with minute yellow dots, and covered with thick greyish-white bloom. Stalk three quarters of an inch long, inserted in a small and rather deep cavity. Flesh greenish-yellow, firm, rich, and sugary. Shoots downy.

A good old plum, suitable either for the dessert or preserving. The tree requires to be grown against an east or a south-east wall; the bloom is very tender and susceptible of early spring frosts.

Bolmar. See *Washington.*

Bolmar's Washington. See *Washington.*

Bonum Magnum. See *White Magnum Bonum*

Bradford Gage. See *Green Gage.*

Bricette. See *Mirabelle Tardive.*

Brignole. See *White Perdrigon.*

Brignole Violette. See *Blue Perdrigon.*

Brugnon Green Gage. See *Green Gage.*

Bullace. See *White Bullace.*

Bury Seedling. See *Coe's Golden Drop.*

Caledonian. See *Goliath.*

De Catalogne. See *White Primordian.*

Catalonian. See *White Primordian.*

Cerisette Blanche. See *White Primordian.*

Chapman's Prince of Wales. See *Prince of Wales.*

Cherry (*Early Scarlet; Miser Plum; Myrobalan; Virginian Cherry*).—Fruit medium sized, cordate, somewhat flattened at the stalk, and terminated at the apex by a small nipple, which bears upon it the remnant of the style like a small bristle. Skin very thick and acrid pale red, and marked with small greyish-white dots. Stalk three quarters of an inch long, slender, and inserted in a small cavity. Flesh yellow, sweet, juicy, and sub-acid, adhering to the stone. Shoots smooth.

More ornamental than useful in the dessert, but is good when baked, or in tarts. Ripe in the beginning or middle of August.

CHESTON (*Dennie; Diaprée Violette; Friars*).—Fruit medium sized, oval, and rather widest at the stalk; suture scarcely discernable. Skin purple, thickly covered with blue bloom. Stalk half an inch long, slender, and not depressed. Flesh deep yellow, firm, brisk, and with a sweet, agreeable flavour, separating from the stone. Shoots downy.

A dessert or preserving plum. Ripe in the beginning and middle of August.

De Chypre. See *Damas Musqué.*

Coe's. See *Coe's Golden Drop.*

COE'S GOLDEN DROP (*Bury Seedling; Coe's; Coe's Imperial; Fair's Golden Drop; Golden Drop; Golden Gage*).—Fruit very large, oval, with a short neck at the stalk, and marked with a deep suture, which extends the whole length of the fruit. Skin pale yellow, marked with a number of dark red spots. Stalk about an inch long, stout, and not depressed. Flesh yellow, rich, sugary, and delicious, adhering closely to the stone. Shoots smooth.

One of the finest plums, and adapted either for the dessert or preserving. It ripens in the end of September. It is much improved by being grown against a wall.

Coe's Imperial. See *Coe's Golden Drop.*

COE'S LATE RED (*St. Martin; St. Martin Rouge*).—Fruit medium sized, round, marked on one side with a deep suture. Skin bright purple, covered with a thin blue bloom. Stalk three quarters of an inch long, not depressed. Flesh yellowish, firm and juicy, with a sweet and sprightly flavour, and separating from the stone. Shoots downy.

As a late plum, ripening in the end of October, and hanging for a month or six weeks later, this is a valuable variety.

COLUMBIA (*Columbia Gage*).—Fruit very large, almost round. Skin deep reddish-purple, dotted with yellowish dots, and thickly covered with blue bloom. Stalk an inch long, inserted in a small, narrow cavity. Flesh orange, with a rich, sugary, and delicious flavour, separating from the stone. Shoots downy. Ripe in the middle of September.

Columbia Gage. See *Columbia.*

Cooper's Blue Gage. See *Blue Gage.*

Cooper's Large (*Cooper's Large American; Cooper's Large Red; La Delicieuse*).—Fruit above medium size, oval, considerably enlarged on one side of the suture, which is broad and shallow. Skin pale yellow on the shaded side, and dark purple on the side next the sun, covered with numerous brown dots. Stalk an inch long, inserted in a small cavity. Flesh yellowish-green, juicy, with a rich and delicious flavour, separating from the stone. Shoots smooth. Ripe in the end of September and beginning of October.

Corse's Admiral.—Fruit large, the size of Red Magnum Bonum; oval, considerably swollen on one side of the suture, which is deep and well defined. Skin light purple, dotted with yellow dots, and covered with pale lilac bloom. Stalk an inch long, inserted in a small cavity. Flesh greenish-yellow, brisk and juicy, pleasantly flavoured, and adhering closely to the stone. Shoots downy.

A preserving plum. Ripe in the end of September.

Corse's Nota Bene. — Fruit large, round. Skin brownish-purple, with somewhat of a greenish tinge on the shaded side, and thickly covered with pale blue bloom. Stalk half an inch long, inserted in a small round cavity. Flesh greenish, firm and juicy, with a rich, sugary flavour, separating from the stone. Shoots smooth.

A dessert plum. Ripe in the middle and end of September.

Damas Blanc. See *White Damask.*

Damas Blanc Gros. See *White Damask.*

Damas Blanc Hâtif Gros. See *White Damask.*

Damas Dronet.—Fruit small, oval, and without any apparent suture. Skin bright green, changing to yellowish as it ripens, covered with a very thin light bloom. Stalk half an inch long, slender, inserted in a narrow and rather deep cavity. Flesh greenish, transparent, firm, very sugary, and separating freely from the stone. Shoots smooth.

A dessert plum. Ripe in the end of August.

Damas d'Italie. See *Italian Damask*.

Damas de Mangeron (*Mangeron*). — Fruit above medium size, round, and inclining to oblate, without any apparent suture. Skin adhering to the flesh, lively purple, strewed with minute yellowish dots, and thickly covered with blue bloom. Stalk half an inch long, slender, inserted in a small cavity. Flesh greenish-yellow, firm, not very juicy, but sugary, and separating from the stone. Shoots smooth.

A baking or preserving plum. Ripe in the beginning and middle of September.

Damas Musqué (*De Chypre; Prune de Malthe*).— Fruit small, roundish, flattened at both ends, and marked with a deep suture. Skin deep purple or nearly black, thickly covered with blue bloom. Stalk half an inch long, inserted in a small cavity. Flesh yellow, firm, very juicy, with a rich and musky flavour, and separating from the stone. Shoots slightly downy.

A dessert or preserving plum. Ripe in the end of August and beginning of September.

Damas de Provence (*Damas de Provence Hâtif*).— Fruit above medium size, roundish, and marked on one side with a deep suture. Skin reddish-purple, covered with blue bloom. Stalk half an inch long, inserted in a small cavity. Flesh yellowish-green, sweet and pleasantly flavoured, separating from the stone. Shoots slightly downy.

A baking plum. Ripe in the end of July and beginning of August.

Damas de Septembre (*Prune de Vacance*).— Fruit small, oval, marked on one side with a distinct suture. Skin brownish-purple, thickly covered with blue bloom. Stalk half an inch long, slender, inserted in a narrow and rather deep cavity. Flesh yellow, firm, rich, and agreeably flavoured when well ripened, and separating from the stone. Shoots downy.

A dessert or preserving plum. Ripe in the end of September.

Damas de Tours. See *Précoce de Tours*.

Damas Vert. See *Green Gage.*

Damaseen. See *Prune Damson.*

Dame Aubert. See *White Magnum Bonum.*

Dame Aubert Blanche. See *White Magnum Bonum.*

Dame Aubert Violette. See *Red Magnum Bonum.*

DAMSON (*Common Damson; Round Damson*).—Fruit very small, roundish-ovate. Skin deep dark purple or black, covered with thin bloom. Flesh greenish-yellow, juicy, very acid, and rather austere till highly ripened, and separating from the stone. Shoots downy.

A well-known preserving plum. Ripe in the end of September.

Dauphine. See *Green Gage.*

La Delicieuse. See *Cooper's Large.*

Dennie. See *Cheston.*

DENNISTON'S SUPERB.—Fruit above medium size, round, and a little flattened, marked with a distinct suture, which extends quite round the fruit. Skin pale yellowish-green, marked with a few purple thin blotches and dots, and covered with bloom. Stalk three quarters of an inch long, inserted in a small cavity. Flesh yellow, firm, not very juicy, but rich, sugary, and vinous, separating from the stone. Shoots downy.

A first-rate dessert plum. Ripe in the middle of August.

DIAMOND.—Fruit very large, oval, marked on one side with a distinct suture, which is deepest towards the stalk. Skin dark purple, approaching to black, and covered with pale blue bloom. Stalk three quarters of an inch long, inserted in a narrow and deep cavity. Flesh deep yellow, coarse in texture, juicy, and with a brisk agreeable acid flavour; it separates with difficulty from the stone. Shoots downy.

One of the best preserving or cooking plums. Ripe in the middle of September.

Diaper. See *Diaprée Rouge.*

DIAPRÉE ROUGE (*Diaper; Imperial Diadem; Mimms; Red Diaper; Roche Corbon*).—Fruit large, obovate. Skin pale red, thickly covered with brown dots, so much so as to make it appear of a dull colour, and covered with thin blue bloom. Stalk half an inch long, inserted in a slight cavity. Flesh greenish-yellow, firm, and fine-grained, separating, but not freely, from the stone, juicy, and of a rich, sugary flavour. Shoots downy.

A good plum for preserving, or the dessert. Ripe in the middle of September.

Diaprée Violette. See *Cheston*.

Downton Imperatrice.—Fruit medium sized, oval, narrowing a little towards the stalk, and slightly marked with a suture on one side. Skin thin and tender, pale yellow. Flesh yellow, separating from the stone, juicy and melting, with a sweet and agreeable subacid flavour. Shoots smooth.

An excellent preserving plum, but only second-rate for the dessert. Ripe in October.

Drap d'Or (*Cloth of Gold; Mirabelle Double; Mirabelle Grosse; Yellow Perdrigon*).—Fruit below medium size, round, indented at the apex, and marked on one side by a distinct but very shallow suture. Skin tender, fine bright yellow, marked with numerous crimson spots, and covered with thin white bloom. Stalk slender, half an inch long, inserted in a small cavity. Flesh yellow, melting, with a rich, sugary flavour, separating from the stone. Shoots downy.

A good dessert plum. Ripe in the middle of August.

Dunmore.—Fruit medium sized, oval. Skin thick, greenish-yellow, becoming of a bright golden yellow when ripe. Stalk half an inch long, inserted in a small cavity. Flesh yellow, tender, juicy, sweet, and richly flavoured, separating from the stone. Shoots smooth.

An excellent dessert plum. Ripe in the end of September and beginning of October.

Early Damask. See *Morocco*.

Early Favorite (*Rivers' Early Favorite; Rivers' No. 1*).—Fruit rather below medium size, roundish-oval, and marked with a shallow suture. Skin deep dark purple, almost black, marked with russet dots, and covered with thin bloom. Flesh greenish-yellow, juicy, sweet, and of excellent flavour, separating from the stone. Shoots smooth.

An excellent early plum, raised by Mr. Rivers, of Sawbridgeworth, from Précoce de Tours. It ripens in the end of July; and is deserving of a wall, when it will ripen in the middle of the month.

Early Morocco. See *Morocco.*

Early Orleans (*Grimwood Early Orleans; Hampton Court; Monsieur Hâtif; Monsieur Hâtif de Montmorency; New Orleans; Wilmot's Early Orleans; Wilmot's Orleans*).—Fruit medium sized, round, flattened at the apex, and marked with a suture, which extends the whole length of one side. Skin deep purple, mottled with darker colour, and covered with thin blue bloom. Stalk slender, about half an inch long, inserted in a rather deep cavity. Flesh yellowish-green, tender, of a rather rich flavour, and separating freely from the stone. Shoots downy.

A second-rate dessert plum, but excellent for culinary purposes. Ripe in the beginning and middle of August.

Early Prolific (*Rivers' Early Prolific; Rivers' No.* 2).—Fruit medium sized, roundish-oval. Skin deep purple, covered with thin bloom. Stalk half an inch long, inserted in a small cavity. Flesh yellowish, juicy, sweet, with a pleasant brisk acidity, separating from the stone. Shoots smooth.

A valuable early plum, ripening in the end of July. The tree is great bearer, and very hardy, rarely ever missing a crop. It was raised by Mr. Rivers, of Sawbridgeworth, from Précoce de Tours.

Early Royal. See *Royale Hâtive.*

Early Russian. See *Quetsche.*

Early Scarlet. See *Cherry.*

Early Yellow. See *White Primordian.*

Egg Plum. See *White Magnum Bonum.*

Emerald Drop.—Fruit medium sized, oval, marked with a deep suture, which is higher on one side than the other. Skin pale yellowish-green. Stalk three quarters of an inch long, inserted in a very shallow cavity. Flesh greenish-yellow, juicy, sweet, and of good flavour, separating from the stone. Shoots smooth.

Ripe in the end of August and beginning of September.

Empress. See *Blue Impératrice.*

Fair's Golden Drop. See *Coe's Golden Drop.*

Fellemberg. See *Italian Quetsche.*

Florence. See *Red Magnum Bonum.*

Flushing Gage. See *Imperial Gage.*

Fonthill. See *Pond's Seedling*.

Fotheringay. See *Fotheringham*.

FOTHERINGHAM (*Fotheringay; Grove House Purple; Red Fotheringham; Sheen*).—Fruit medium sized, obovate, with a well-defined suture, which is higher on one side than the other. Skin deep reddish-purple on the side next the sun, and bright red where shaded, covered with thin blue bloom. Stalk an inch long, not deeply inserted. Flesh pale greenish-yellow, not juicy, sugary, with a pleasant subacid flavour, and separating from the stone. Shoots smooth.

A good dessert plum. Ripe in the middle of August.

Franklin. See *Washington*.

Friar's. See *Cheston*.

FROST GAGE (*American Damson; Frost Plum*).—Fruit small, roundish-oval, and marked with a distinct suture. Skin deep purple, strewed with russet dots, and covered with a thin bloom. Stalk about three quarters of an inch long. Flesh greenish-yellow, juicy, sweet, and rather richly flavoured, adhering to the stone. Shoots smooth.

An excellent little plum. Ripe in October. The tree is a great bearer.

Frost Plum. See *Frost Gage*.

GENERAL HAND.—Fruit very large, roundish-oval, marked with a slight suture. Skin deep golden yellow, marbled with greenish-yellow. Stalk long, inserted in a shallow cavity. Flesh pale yellow, coarse, not very juicy, sweet, and of a good flavour, and separating from the stone. Shoots smooth.

A preserving plum. Ripe in the beginning and middle of September.

German Gage. See *Bleeker's Yellow Gage*.

German Prune. See *Quetsche*.

German Quetsche. See *Quetsche*.

GISBORNE'S (*Gisborne's Early; Paterson's*).—Fruit rather below medium size, roundish-oval, marked with a distinct suture. Skin greenish-yellow, but changing as it ripens to fine amber, with a few crimson spots, and numerous grey russet dots interpersed. Stalk half an inch to three quarters long, inserted in a very shallow cavity. Flesh yellow, firm, coarse-grained, and not very juicy, briskly acid, with a slight sweetness, and separating from the stone. Shoots downy.

A cooking plum. Ripe in the middle of August. The tree is an early and abundant bearer.

Gisborne's Early. See *Gisborne's.*

Gloire de New York. See *Hulings' Superb.*

Golden Drop. See *Coe's Golden Drop.*

Golden Gage. See *Coe's Golden Drop.*

GOLIATH (*Caledonian; St. Cloud; Steers' Emperor; Wilmot's Late Orleans*).—Fruit large, oblong, with a well-marked suture, one side of which is higher than the other. Skin deep reddish-purple, but paler on the shaded side, and covered with thin blue bloom. Stalk three quarters of an inch long, inserted in a deep cavity. Flesh yellow, juicy, brisk, and of good flavour, adhering to the stone. Shoots downy.

A fine showy plum, and though only of second-rate quality for the dessert, is excellent for preserving and other culinary purposes. Ripe in the end of August. This is sometimes, but erroneously, called *Nectarine Plum.*

Gonne's Green Gage. See *Yellow Gage.*

Great Damask. See *Green Gage.*

GREEN GAGE (*Abricot Vert; Bradford Green Gage; Brugnon Green Gage; Damas Vert; Dauphine; Great Green Damask; Grosse Reine; Ida Green Gage; Isleworth Green Gage; Mirabelle Vert Double; Queen Claudia; Reine Claude; Reine Claude Grosse; Rensselaar Gage; Schuyler Gage; Sucrin Vert; Tromphe Garçon; Trompe Valet; Verdacia; Verdochio; Vert Bonne; Verte Tiquetée; Wilmot's Green Gage*).—Fruit medium sized, round, and a little flattened at both ends; dimpled at the apex, and marked on one side by a shallow suture, which extends from the stalk to the apex. Skin tender, yellowish-green, but, when fully ripe, becoming of a deeper yellow, cloudedwith green, and marked with crimson spots, and covered with thin ashy-grey bloom. Stalk half an inch to three quarters long, inserted in a small cavity. Flesh greenish-yellow, tender, melting, and very juicy, with a rich, sugary, and most delicious flavour; it separates freely from the stone. Shoots smooth.

One of the richest of all the plums. Ripe in the middle and end of August.

Grimwood's Early Orleans. See *Early Orleans.*

Grosse Luisante. See *White Magnum Bonum.*

Grosse Reine. See *Green Gage.*

Grosse Rouge de Septembre. See *Belle de Septembre.*

Grove House Purple. See *Fotheringham.*

GUTHRIE'S APRICOT. — Fruit above medium size, roundish-oval. Skin yellow, strewed with crimson dots, and covered with thin bloom. Stalk rather long, set in a small depression. Flesh yellow, rather coarse, juicy and sweet, adhering to the stone. Shoots smooth.

A second-rate dessert plum. Ripe in the end of August.

Guthrie's Aunt Ann. See *Aunt Ann.*

GUTHRIE'S LATE GREEN.—Fruit above medium size, round, marked with a suture, which is swollen on one side. Skin yellow, clouded with green, and covered with a thin bloom. Stalk three quarters of an inch long, inserted in a small cavity. Flesh yellow, firm, not very juicy, but exceedingly rich and sugary, adhering slightly to the stone. Shoots smooth.

A very fine dessert plum, rivalling the Green Gage, and ripening about a month later—the end of September. The tree is hardy, and a good bearer.

Hampton Court. See *Early Orleans.*

Howell's Large. See *Nectarine.*

HULINGS' SUPERB (*Gloire de New York; Keyser's Plum*).—Fruit very large, roundish-oval, marked with a shallow suture. Skin greenish-yellow, covered with a thin bloom. Stalk short and stout, inserted in a small round cavity. Flesh greenish-yellow, rather coarse, but rich and sugary, and with a fine brisk flavour; it adheres to the stone. Shoots downy.

A fine, large, and richly-flavoured plum. Ripe in the end of August.

ICKWORTH'S IMPÉRATRICE (*Knight's No.* 6). — Fruit large, obovate. Skin purple, marked with yellow streaks. Stalk stout, an inch or more in length. Flesh greenish-yellow, tender and juicy, with a rich, sugary flavour, and adhering to the stone. Shoots smooth.

An excellent late dessert plum. Ripe in October. It will hang till it shrivels, and is then very rich in flavour.

Impératrice. See *Blue Impératrice.*

Impératrice Blanche. See *White Impératrice.*

Imperial Gage (*Flushing Gage; Prince's Imperial Gage*).—Fruit above medium size, oval, marked with a distinct suture. Skin greenish-yellow, marked with green stripes, and covered with thick bloom. Stalk an inch long, inserted in a small, even cavity. Flesh greenish, tender, melting, and very juicy, with a rich and brisk flavour, separating from the stone. Shoots slightly downy.

A dessert plum. Ripe in the middle of September.

Imperial Diadem. See *Diaprée Rouge.*

Imperial Ottoman. — Fruit below medium size, roundish. Skin dull yellow, covered with a thin bloom. Stalk slender, curved, three quarters of an inch long, inserted in a slight cavity. Flesh melting, juicy, and sweet, adhering to the stone. Shoots slightly downy.

An early dessert plum, ripening in the beginning of August.

Impériale. See *Red Magnum Bonum.*

Impériale Blanche. See *White Magnum Bonum.*

Impériale de Milan (*Prune de Milan*).—Fruit large, oval, somewhat flattened on one side, where it is marked with a rather deep suture extending the whole length of the fruit. Skin dark purple, streaked and dotted with yellow, and covered with thick blue bloom. Stalk about an inch long, inserted in a narrow and rather deep cavity. Flesh yellowish, firm and juicy, richly flavoured and sweet, with a slight musky aroma, and adhering to the stone. Shoots smooth.

An excellent late dessert and preserving plum. Ripe in the beginning of October.

Impériale Rouge. See *Red Magnum Bonum.*

Impériale Violette. See *Red Magnum Bonum.*

Irving's Bolmar. See *Washington.*

Isabella.—Fruit medium sized, obovate. Skin deep dull red, but paler red where shaded, and strewed with darker red dots. Stalk three quarters of an inch long. Flesh yellow, juicy, rich, and adhering to the stone. Shoots downy.

A dessert and preserving plum. Ripe in the beginning of September.

Isleworth Green Gage. See *Green Gage.*

Italian Damask (*Damas d'Italie*).—Fruit medium sized, roundish, slightly flattened at the base, and marked with a well-defined suture. Skin thick, membranous, and rather bitter, of a pale purple colour, changing to brownish as it ripens, and covered with fine blue bloom. Stalk three quarters of an inch long, slender, inserted in a deep cavity. Flesh yellowish-green, firm, rich, sugary, and excellent, separating from the stone. Shoots smooth.

A dessert and preserving plum. Ripe in the beginning of September.

Italian Quetsche (*Altesse Double; Fellemberg; Quetsche d'Italie; Prune d'Italie; Semiana*).—Fruit large, oval, narrowing a little towards the stalk, and marked with a shallow suture. Skin dark purplish-blue, marked with yellow dots, and covered with thick blue bloom. Stalk half an inch long, stout, and inserted in a pretty deep cavity. Flesh greenish-yellow, firm, not very juicy, sweet, and richly flavoured; when highly ripened separating from the stone. Shoots smooth.

An excellent dessert or preserving plum. Ripe in the beginning of September, and will hang till it shrivels, when it is very rich and delicious. This, I believe, to be the true *Semiana*. It well deserves a wall.

Jaune de Catalogue. See *White Primordian.*

Jaune Hâtive. See *White Primordian.*

Jefferson.—Fruit large, oval, narrowing a little towards the stalk, and marked with a very faint suture. Skin greenish-yellow, becoming of a rich golden yellow, flushed with red on the side next the sun, and dotted with red dots. Stalk an inch long, thin, and inserted in a shallow cavity. Flesh yellow, firm and juicy, rich, sugary, and delicious, separating from the stone. Shoots smooth.

A richly-flavoured dessert plum. Ripe in the beginning and middle of September.

Jenkins' Imperial. See *Nectarine.*

July Green Gage (*Reine Claude Hâtive*).—Fruit the size and shape of the Green Gage. Skin thin, of a fine deep yellow colour, flushed with bright crimson on the side next the sun, and strewed with darker crimson dots; the whole covered with a delicate white bloom. Stalk three quarters of an inch long, slightly depressed. Flesh deep yellow, very tender and juicy, sugary, and richly flavoured, separating from the stone. Shoots smooth.

A first-rate and most delicious early plum, equal in all respects to the Green Gage, and ripening in the end of July.

Keyser's Plum. See *Hulings' Superb.*

KIRKE'S.—Fruit above medium size, round, and marked with a very faint suture. Skin dark purple, with a few yellow dots, and covered with a thick blue bloom. Stalk three quarters of an inch long, inserted in a slight depression. Flesh greenish-yellow, firm, juicy, sugary, and very richly flavoured, separating from the stone. Shoots smooth.

A delicious dessert plum. Ripe in the beginning and middle of September.

Kirke's Stoneless. See *Stoneless.*

Knevett's Late Orleans. See *Nelson's Victory.*

KNIGHT'S GREEN DRYING (*Large Green Drying*).—Fruit large, round, and marked with a shallow suture. Skin greenish-yellow, and covered with thin white bloom. Flesh yellowish, firm, not very juicy, sugary, and richly flavoured when highly ripened; adhering to the stone. Shoots smooth.

A dessert plum. Ripe in the middle and end of September; and succeeds best against a wall.

Knight's No. 6. See *Ickworth Impératrice.*

Large Green Drying. See *Knight's Green Drying.*

LATE GREEN GAGE (*Reine Claude d'October; Reine Claude Tardive*).—Fruit of the same shape but smaller than the Green Gage. Skin greenish-yellow, covered with thin white bloom. Stalk stout, three quarters of an inch long. Flesh green, juicy, rich and sugary, separating from the stone. Shoots smooth.

A dessert plum. Ripe in the end of September and beginning of October.

LATE ORLEANS (*Monsieur Tardive; Black Orleans*).—Fruit very similar in appearance to the Orleans, but larger. The flesh is more richly flavoured and sugary. Shoots smooth.

A valuable late dessert plum. Ripe in the end of September and beginning of October, and will hang till November.

Lawrence's Favorite. See *Lawrence Gage.*

LAWRENCE GAGE (*Lawrence's Favorite*).—Fruit large, round, and flattened at both ends. Skin dull yellowish-green, streaked with darker green on the side exposed to the sun, veined with brown, and covered all over with thin grey bloom. Stalk half an inch long, inserted in a narrow cavity. Flesh greenish, tender, melting, and juicy, rich, sugary, and with a fine vinous, brisk flavour, separating from the stone. Shoots downy.

A delicious dessert plum. Ripe in the beginning of September.

Leipzig. See *Quetsche.*

Little Queen Claude. See *Yellow Gage.*

LOMBARD (*Bleeker's Scarlet; Beckman's Scarlet*).—Fruit medium size, roundish-oval, and marked with a shallow suture. Skin purplish-red, dotted with darker red, and covered with thin bloom. Stalk half an inch long, slender, set in a wide funnel-shaped cavity. Flesh yellow, juicy, and pleasantly flavoured, adhering to the stone. Shoots smooth.

A preserving or culinary plum. Ripe in the end of August and beginning of September.

London Plum. See *White Primordian.*

Long Damson. See *Prune Damson.*

LUCOMBE'S NONESUCH.—Fruit above medium size, round, and compressed on the side, where it is marked with a broad suture. Skin greenish-yellow, streaked with orange, and covered with a greyish-white bloom. Stalk three quarters of an inch long, inserted in a rather wide cavity. Flesh greenish-yellow, firm, juicy, rich, and sugary, with a pleasant briskness, and adhering to the stone. Shoots smooth.

A dessert and preserving plum, bearing considerable resemblance to the Green Gage, but not so richly flavoured. Ripe in the end of August.

MCLAUGHLIN.—Fruit large, roundish-oblate. Skin thin and tender, of a fine yellow colour, dotted and mottled with red, and covered with thin grey bloom. Stalk three quarters of an inch long, inserted in a small round cavity. Flesh yellow, firm, very juicy, sweet, with a rich luscious flavour, and adhering to the stone. Shoots smooth.

An excellent plum, ripening in the end of August.

Maître Claude. See *White Perdrigon.*

MAMELONNÉ (*Mamelon Sageret*).—Fruit medium sized, roundish-oval, tapering with a pear-shaped neck towards the stalk, and frequently furnished with a nipple at the apex. Skin yellowish-green, mottled with red next the sun, and covered with grey bloom. Stalk short, inserted without depression. Flesh yellowish, firm, very juicy, sugary, and richly flavoured, separating freely from the stone. Shoots smooth.

An excellent dessert plum, ripening about the middle of August.

Mimms. See *Diaprée Rouge.*

MIRABELLE PETITE (*Mirabelle; Mirabelle Blanche; Mirabelle Jaune; Mirabelle Perle; Mirabelle de Vienne; White Mirabelle*).—Fruit produced in clusters, small, roundish-oval, and marked with a faint suture on one side. Skin of a fine yellow colour, sometimes marked with crimson spots on the side exposed to the sun, and covered with thin white bloom. Stalk three quarters of an inch long, inserted without depression. Flesh deep yellow, firm, pretty juicy, sweet, and briskly flavoured, separating from the stone. Shoots downy.

A valuable little plum for preserving, and all culinary purposes. Ripe in the middle of August. The tree forms a handsome pyramid, and is a most abundant bearer.

Mirabelle Blanche. See *Mirabelle Petite.*

Mirabelle Double. See *Drap d'Or.*

Mirabelle Grosse. See *Drap d'Or.*

Mirabelle Jaune. See *Mirabelle Petite.*

Mirabelle d'Octobre. See *Mirabelle Tardive.*

Mirabelle Perle. See *Mirabelle Petite.*

MIRABELLE TARDIVE (*Bricette; Mirabelle d'Octobre; Petite Bricette*).—Fruit small, roundish-oval, sometimes quite round, and marked with a distinct suture. Skin thin and tender, yellowish-white, dotted and speckled with red, and covered with thin white bloom. Stalk half an inch long, slender, inserted in a shallow and narrow cavity. Flesh yellowish-white, firm, very juicy, with a brisk vinous flavour, and adhering partially to the stone. Shoots smooth.

An excellent preserving and culinary plum. Ripe in

October. The tree forms a handsome pyramid, and is an excellent bearer.

Mirabelle de Vienne. See *Mirabelle Petite.*

Mirabelle Vert Double. See *Green Gage.*

Miser Plum. See *Cherry.*

Miriam. See *Royale Hâtive.*

Mogul Rouge. See *Red Magnum Bonum.*

Monsieur. See *Orleans.*

Monsieur à Fruits Jaune. See *Yellow Orleans.*

Monsieur Hâtif. See *Early Orleans.*

Monsieur Hâtif de Montmorency. See *Early Orleans.*

Monsieur Ordinaire. See *Orleans.*

Monsieur Tardive. See *Late Orleans.*

Monsieur Tardive. See *Suisse.*

Monstrueuse de Bavay. See *Reine Claude de Bavay.*

De Montfort.—Fruit medium sized, roundish, inclining to ovate, with a well-marked suture on one side. Skin dark purple, covered with a thin pale blue bloom. Stalk half an inch long, not deeply inserted. Flesh greenish-yellow, tender and melting, with a thick syrupy and honied juice, and when it hangs till it shrivels is quite a sweetmeat; separates from the stone, which is small. Shoots smooth.

A delicious dessert plum. Ripe in the middle of August. It bears considerable resemblance to Royale Hâtive, but is larger, and appears to be an improved form of that variety.

Morocco (*Black Damask; Black Morocco; Early Damask; Early Morocco*).—Fruit medium sized, roundish, flattened at the apex, and marked on one side with a shallow suture. Skin very dark purple, almost black, and covered with thin pale blue bloom. Stalk stout, about half an inch long. Flesh greenish-yellow, juicy, with a sweet, brisk flavour, and slightly adhering to the stone. Shoots downy.

An excellent early plum. Ripe in the beginning of August.

Myrobalan. See *Cherry.*

Nectarine (*Howell's Large; Jenkins' Imperial;*

*Peach; Prune Pêche*).—Fruit large, roundish, and handsomely formed. Skin purple, covered with fine azure bloom. Stalk half an inch long, stout, inserted in a wide and shallow cavity. Flesh dull greenish-yellow, with a sweet and brisk flavour, separating from the stone. Shoots smooth.

A good plum for preserving and other culinary purposes. Ripe in the middle of August. This is quite distinct from the Goliath, which is sometimes called by the same name.

NELSON'S VICTORY (*Knevett's Late Orleans*).—Fruit medium sized, round, and marked with a shallow suture. Skin deep purple, and covered with blue bloom. Stalk half an inch long, set in a shallow cavity. Flesh firm, rather coarse, sweet and briskly flavoured, adhering to the stone. Shoots smooth.

A culinary plum. Ripe in the middle of September. The tree is a very abundant bearer.

New Orleans. See *Early Orleans*.

Noire Hâtive. See *Précoce de Tours*.

Œuf Rouge. See *Red Magnum Bonum*.

Old Apricot. See *Apricot*.

ORLEANS (*Anglaise Noire; Monsieur; Monsieur Ordinaire; Prune d'Orleans; Red Damask*).—Fruit medium sized, round, somewhat flattened at the ends, and marked with a suture, which is generally higher on one side than the other. Skin tender, dark red, becoming purple when highly ripened, and covered with blue bloom. Stalk three quarters of an inch long, inserted in a considerable depression. Flesh yellowish, tender, sweet, and briskly flavoured, separating from the stone. Shoots downy.

A preserving and culinary plum. Ripe in the middle and end of August.

Parker's Mammoth. See *Washington*.

Paterson's. See *Gisborne's*.

PEACH.—Fruit large, roundish, inclining to oblate, marked with a shallow suture on one side. Skin bright red, dotted with amber. Flesh tender, melting, juicy, very sweet and luscious, separating freely from the stone. Shoots smooth.

An early dessert plum. Ripe in the beginning of August. It is quite distinct from the Nectarine Plum, which is also known by this name; and was introduced some years ago by Mr. Rivers, of Sawbridgeworth.

Peach. See *Nectarine.*

Perdrigon Blanc. See *White Perdrigon*

Perdrigon Rouge. See *Red Perdrigon.*

Perdrigon Violet. See *Blue Perdrigon.*

PERDRIGON VIOLET HÂTIF.—Fruit medium sized, roundish-oval. Skin purple. Flesh rich, juicy, and excellent, separating from the stone. Shoots downy.

A first-rate dessert plum. Ripe in the middle of August. The tree is very hardy, and an abundant bearer. This is not the same as Perdrigon Hâtif and Moyeu de Bourgogne with which it is made synonymous in the Horticultural Society's Catalogue, both of these being yellow plums.

Petite Bricette. See *Mirabelle Tardive.*

Petite Damas Vert. See *Yellow Gage.*

Pickett's July. See *White Primordian.*

Pigeon's Heart. See *Queen Mother.*

Pond's Purple. See *Pond's Seedling.*

POND'S SEEDLING (*Fonthill; Pond's Purple*).—Fruit very large, oval, widest at the apex and narrowing towards the stalk, marked with a wide suture. Skin fine dark red, thickly strewed with grey dots, and covered with thin bluish bloom. Stalk three quarters of an inch long, inserted without depression. Flesh yellowish, rayed with white, juicy, and briskly flavoured, adhering to the stone. Shoots smooth.

A valuable culinary plum. Ripe in the beginning and middle of September.

PRÉCOCE DE BERGTHOLD.—This is a small, roundish-oval plum of a yellow colour, similar in appearance to, but of better flavour than, White Primordian. The flesh is juicy and sweet, separating from the stone. Shoots downy. It is very early, ripening before the White Primordian in the latter end of July.

PRÉCOCE DE TOURS (*Damas de Tours; Noire Hâtive; Prune de Gaillon; Violette de Tours*).—Fruit below medium size, oval, sometimes inclining to obovate, and marked with a shallow indistinct suture. Skin deep purple, almost black, thickly covered with blue bloom. Stalk half an inch long, slender, inserted in a very slight depression. Flesh dull yellow, rather juicy and sweet, with a pleasant flavour, and adhering closely to the stone. Shoots downy.

A second-rate dessert plum, but well adapted for culinary use. Ripe in the beginning of August.

Prince Englebert.—Fruit very large, oval, and marked with a shallow suture. Skin of a uniform deep purple, covered with minute russety dots, the whole thickly covered with pale grey bloom. Stalk half an inch long, inserted in a rather deep cavity. Flesh yellow, rather firm, sweet, juicy, with a brisk and rich flavour, and adhering to the stone. Shoots smooth.

An excellent plum either for the dessert or for culinary purpose, and "delicious when preserved." Ripe in September. The tree is a great bearer.

Prince of Wales (*Chapman's Prince of Wales*).—Fruit above medium size, roundish, inclining to oval, marked with a distinct suture. Skin bright purple, covered with thick azure bloom, and dotted with yellow dots. Stalk short and stout, inserted in a slight cavity. Flesh coarse-grained, yellowish, juicy, and sweet, with a brisk flavour, and separating from the stone. Shoots smooth.

A dessert plum of second-rate quality, but suitable for all culinary purposes. Ripe in the beginning of September. The tree is a very abundant bearer.

Prince's Imperial Gage. See *Imperial Gage.*

Prune d'Allemagne. See *Quetsche.*

Prune d'Ast. See *d'Agen.*

Prune Damson (*Damascene; Long Damson; Shropshire Damson*).—The fruit of this variety is much larger than that of the common Black Damson, and more fleshy. It is generally preferred for preserving, and of all the other Damsons makes the best jam. The flesh adheres to the stone. Shoots downy.

The tree is not such a good bearer as the common Damson. Ripe in the middle of September.

Prune de Gaillon. See *Précoce dé Tours.*

Prune d'Italie. See *Italian Quetsche.*

Prune de Milan. See *Impériale de Milan.*

Prune d'Orleans. See *Orleans.*

Prune Pêche. See *Nectarine.*

Prune Pêche. See *Peach.*

Prune du Roi. See *d'Agen.*

Purple Egg. See *Red Magnum Bonum.*

PURPLE GAGE (*Reine Claude Violette; Violet Gage*).—Fruit medium sized, round, slightly flattened at the ends, and marked with a shallow suture. Skin fine light purple, dotted with yellow, and covered with pale blue bloom. Stalk an inch long, inserted in a small cavity. Flesh greenish-yellow, firm, with a rich, sugary, and most delicious flavour, and separating from the stone. Shoots smooth.

A dessert plum of the greatest excellence, and particularly richly flavoured if allowed to hang till it shrivels. Ripe in the beginning of September.

Queen Claudia. See *Green Gage.*

QUEEN MOTHER (*Pigeon's Heart*). — Fruit below medium size, round, and marked with a slight suture. Skin dark red next the sun, but paler towards the shaded side, where it is yellow, and covered all over with reddish dots. Stalk half an inch long, inserted in a small depression. Flesh yellow, rich, and sugary, separating from the stone. Shoots smooth.

A dessert plum. Ripe in September.

QUETSCHE (*Early Russian; German Prune; German Quetsche; Leipzig; Prune d'Allemagne; Sweet Prune; Turkish Quetsche; Zwetsche*).—Fruit medium sized, oval, narrowing towards the stalk, flattened on one side, where it is marked with a distinct suture. Skin dark purple, dotted with grey dots and veins of russet, and covered with blue bloom. Stalk an inch long. Flesh firm, juicy, sweet, and brisk, separating from the stone. Shoots smooth.

A culinary plum. Ripe in the end of September.

Quetsche d'Italie. See *Italian Quetsche.*

Red Damask. See *Orleans.*

Red Diaper. See *Diaprée Rouge.*

Red Fotheringham. See *Fotheringham.*

Red Imperial. See *Red Magnum Bonum.*

RED MAGNUM BONUM (*Askew's Purple Egg; Dame Aubert Violette; Florence; Impériale; Impériale Rouge; Impériale Violette; Mogul Rouge; Œuf Rouge; Purple Egg; Red Egg; Red Imperial*).—Fruit large, oval, and narrowing a little towards the stalk; marked with a distinct suture, one side of which is frequently higher than the other. Skin deep red where exposed to the sun, but paler in the shade; strewed with grey dots, and covered with blue bloom. Stalk an inch long, inserted in a small cavity. Flesh greenish, firm, rather coarse, not very juicy, and briskly flavoured, and separating from the stone. Shoots smooth.

A culinary plum. Ripe in the beginning and middle of September.

RED PERDRIGON (*Perdrigon Rouge*).—Fruit small, roundish-oval. Skin fine deep red, marked with fawn-coloured dots, and thickly covered with pale blue bloom. Stalk an inch long, stout, inserted in a round cavity. Flesh clear yellow, firm, rich, juicy, and sugary, and separating from the stone. Shoots downy.

A dessert plum. Ripe in the middle and end of September.

Reina Nova. See *Belle de Septembre.*

Reine Claude. See *Green Gage.*

REINE CLAUDE DE BAVAY (*Monstrueuese de Bavay*).—Fruit large, roundish, and flattened at both ends. Skin greenish-yellow, mottled and streaked with green, and covered with a delicate white bloom. Stalk half an inch long, inserted in a small cavity. Flesh yellow, tender, melting, and very juicy, with a rich, sugary flavour, and separating from the stone. Shoots smooth.

A first-rate dessert plum of exquisite flavour. Ripe in the end of September and beginning of October.

Reine Claude Diaphane. See *Transparent Gage.*

Reine Claude Grosse. See *Green Gage.*

Reine Claude d'Octobre. See *Late Green Gage.*

Reine Claude Petite. See *Yellow Gage.*

REINE CLAUDE ROUGE (*Reine Claude Rouge Van Mons*).—Fruit very large, roundish-oval. Skin reddish-purple, dotted with yellow russet dots, and covered with very thick bluish-white bloom. Stalk thick, about an inch long, inserted in a deep cavity. Flesh tender, juicy, sugary, and deliciously flavoured, and separating from the stone. Shoots downy.

An excellent dessert plum. Ripe in the end of August and beginning of September.

Reine Claude Tardive. See *Late Green Gage.*

Reine Claude Violette. See *Purple Gage.*

Rensselaer Green Gage. See *Green Gage.*

Robe de Sargent. See *d'Agen.*

Roche Corbon. See *Diaprée Rouge.*

Roe's Autumn Gage. See *Autumn Gage.*

Rotherham. See *Winesour.*

Round Damson. See *Damson.*

ROYAL DAUPHINE.—Fruit large, oval. Skin pale red on the shaded side, marked with green specks, but darker red next the sun; mottled with darker and lighter shades, and covered with violet bloom. Stalk an inch long, stout. Flesh greenish-yellow, sweet, and subacid, separating from the stone. Shoots smooth.

A culinary plum. Ripe in the beginning of September.

Royal. See *Royale.*

Royal Red. See *Royale.*

ROYALE (*Royal; Royal Red; Sir Charles Worsley's*). —Fruit rather above medium size, round, narrowing a little towards the stalk, marked with a distinct suture Skin light purple, strewed with fawn-coloured dots, and covered with thick pale blue bloom. Stalk about an inch long, stout, and inserted in a small cavity. Flesh yellowish, firm, melting, and juicy, with a rich, delicious flavour, and separating from the stone. Shoots downy.

A dessert plum of first-rate quality. Ripe in the middle of August.

ROYALE HÂTIVE (*Early Royal; Miviam*).—Fruit medium sized, roundish, narrowing towards the apex. Skin light purple, strewed with fawn-coloured dots, and covered with blue bloom. Stalk half an inch long, stout, and inserted without depression. Flesh yellow, juicy and melting, with an exceedingly rich and delicious flavour, and separating from the stone. Shoots downy.

A first-rate dessert plum. Ripe in the beginning and middle of August.

Royale de Tours.—Fruit large, roundish, flattened at the apex, and marked with a distinct suture. Skin light purple, strewed with small yellow dots, and covered with thick blue bloom. Stalk three quarters of an inch long, inserted in a small cavity. Flesh greenish-yellow, tender, very juicy, and richly flavoured, separating from the stone. Shoots downy.

An excellent plum either for the dessert or for preserving. Ripe in the middle of August.

St. Barnabe. See *White Primordian.*

St. Catherine.—Fruit medium sized, obovate, tapering towards the stalk, and marked with a suture which is deepest at the stalk. Skin pale yellow, dotted with red, and covered with pale bloom. Stalk three quarters of an inch long, slender, and inserted in a narrow cavity Flesh yellow, tender and melting, rich, sugary, and briskly flavoured, adhering to the stone. Shoots smooth.

A dessert and preserving plum. Ripe in the middle of September.

St. Cloud. See *Goliath.*

St. Etienne.—Fruit medium sized, roundish-oval, frequently somewhat heart-shaped. Skin thin, greenish-yellow, strewed with red dots and flakes, and sometimes with a red blush on the side next the sun. Stalk half an inch long, inserted in a narrow cavity. Flesh yellow, tender, melting and juicy, rich and delicious, separating from the stone. Shoots smooth.

A first-rate dessert plum. Ripe in the beginning and middle of August.

St. Martin. See *Coe's Late Red.*

St. Martin Rouge. See *Coe's Late Red.*

St. Martin's Quetsche.—Fruit medium sized, ovate, or rather heart-shaped. Skin pale yellow, covered with white bloom. Flesh yellowish, sweet, and well-flavoured, separating from the stone. Shoots smooth.

A very late plum. Ripe in the middle of October.

St. Maurin. See *d'Agen.*

Sans Noyau. See *Stoneless.*

Schuyler Gage. See *Green Gage.*

Semiana. See *Italian Quetsche.*

Shailer's White Damson. See *White Damson.*

Sharp's Emperor. See *Victoria*.

Sheen. See *Fotheringham*.

Shropshire Damson. See *Prune Damson*.

Sir Charles Worsley's. See *Royale*.

SMITH'S ORLEANS.—Fruit large, oval, or roundish-oval, widest towards the stalk, and marked with a deep suture. Skin reddish-purple, strewed with yellow dots, and covered with thick blue bloom. Stalk half an inch long, slender, inserted in a deep cavity. Flesh deep yellow, firm, juicy, richly briskly flavoured, and perfumed, adhering to the stone. Shoots smooth.

An excellent plum. Ripe in the end of August.

STANDARD OF ENGLAND.—Fruit above medium size, obovate, and marked with a shallow suture. Skin pale red, strewed with yellow dots, and covered with thin bloom. Stalk three quarters of an inch long, inserted in a small cavity. Skin rather firm, juicy, and briskly flavoured, separating from the stone. Shoots smooth.

A culinary plum. Ripe in the beginning of September.

Steer's Emperor. See *Goliath*.

STONELESS (*Kirke's Stoneless; Sans Noyau*).—Fruit small, oval. Skin dark purple, or rather black, covered with blue bloom. Stalk half an inch long. Flesh greenish-yellow, at first harsh and acid, but when highly ripened and when it begins to shrivel it is mellow and agreeable. The kernel is not surrounded by any bony deposit. Shoots downy. Ripe in the beginning of September.

Sucrin Vert. See *Green Gage*.

SUISSE (*Monsieur Tardive; Switzer's Plum*).—Fruit medium sized, round, slightly depressed at the apex, and marked with a very shallow suture. Skin of a fine dark purple next the sun, but paler on the shaded side, strewed with yellow dots, and covered with blue bloom. Stalk three quarters of an inch long, inserted in a rather wide cavity. Flesh greenish-yellow, juicy and melting, with a rich, brisk flavour, and adhering to the stone. Shoots smooth.

A preserving plum. Ripe in the beginning of October.

Sweet Prune. See *Quetsche*.

Switzer's Plum. See *Suisse*.

TARDIVE DE CHALONS.—Fruit rather small, round, inclining to oval, and marked with a well-defined suture.

Skin brownish-red, thinly strewed with minute dots. Stalk three quarters of an inch long. Flesh, firm, tender, sweet, and well flavoured, separating with difficulty from the stone. Shoots downy.

A dessert or preserving plum. Ripe in October.

Topaz (*Guthrie's Topaz*).—Fruit medium sized, oval, narrowing at the stalk, and marked with a distinct suture. Skin fine clear yellow, covered with thin bloom. Stalk an inch long, inserted in a small cavity. Flesh yellow, juicy, sweet, and richly flavoured, adhering to the stone. Shoots smooth.

A dessert plum, ripening in the middle and end of September, and hanging till it shrivels.

Transparent Gage (*Prune Transparente; Reine Claude Diaphane*).—Fruit rather larger than the Green Gage, roundish-oval, marked with a shallow suture. Skin thin and so transparent as to show the texture of the flesh, and also the stone when the fruit is held up between the eye and the light; pale yellow, dotted and marbled with red. Stalk three quarters of an inch long, thin, and inserted in a shallow cavity. Flesh yellow, rather firm and transparent, very juicy, and with a rich honied sweetness, separating with difficulty from the stone. Shoots smooth.

A most delicious dessert plum. Ripe in the beginning of September.

Trompe Garçon. See *Green Gage*.

Trompe Valet. See *Green Gage*.

Turkish Quetsche. See *Quetsche*.

Verdacia. See *Green Gage*.

Verdochio. See *Green Gage*.

Verte Bonne. See *Green Gage*.

Verte Tiquetée. See *Green Gage*.

Victoria (*Alderton; Denyer's Victoria; Sharp's Emperor*).—Fruit large, roundish-oval, marked with a shallow suture. Skin bright red on the side next the sun, but pale red on the shaded side, and covered with thin bloom. Stalk three quarters of an inch long, stout. Flesh yellow, very juicy, sweet, and pleasantly flavoured, separating from the stone. Shoots downy.

A culinary plum. Ripe in the beginning and middle of September.

VIOLET DAMASK (*Damas Violet*).—Fruit medium sized, oval, narrowing towards the stalk, and slightly flattened on one side. Skin reddish-purple, covered with delicate blue bloom. Stalk half an inch long. Flesh yellow, firm, sweet, and briskly flavoured, separating from the stone. Shoots downy.

A dessert or preserving plum. Ripe in the end of August.

Violet Gage. See *Purple Gage.*

Violet Perdrigon. See *Blue Perdrigon.*

Violette de Tours. See *Précoce de Tours.*

Virginian Cherry. See *Cherry.*

WASHINGTON (*Bolmar; Bolmar's Washington; Franklin; Irving's Bolmar; Parker's Mammoth*).—Fruit large, roundish-ovate, with a faint suture on one side. Skin of a fine deep yellow, marked with crimson dots, and covered with grey bloom. Stalk three quarters of an inch long, inserted in a wide and shallow cavity. Flesh yellow, firm, juicy, sweet, and pleasantly flavoured, separating from the stone. Shoots downy.

A handsome plum, suitable for the dessert, but better adapted for preserving. Ripe in the middle of September.

Wentworth. See *White Magnum Bonum.*

WHITE BULLACE (*Bullace*).—Fruit small, round. Skin pale yellowish-white, mottled with red on the side next the sun. Flesh firm, juicy, sweet, and subacid, adhering to the stone. Shoots downy.

A culinary plum. Ripe in October.

WHITE DAMASK (*Damas Blanc; Damas Blanc Gros; Damas Blanc Hâtif Gros*).—Fruit rather below medium size, roundish, inclining to oval, and swollen on one side of the suture. Skin greenish-yellow, covered with white bloom. Stalk half an inch long, stout. Flesh sweet, pleasantly flavoured, and separating from the stone. Shoots smooth.

A culinary plum. Ripe in the beginning of September.

WHITE DAMSON (*Shailer's White Damson*).—Fruit small, oval. Skin pale yellow, covered with thin white bloom. Stalk half an inch long, slender. Flesh yellow, sweet, and agreeably acid, adhering to the stone. Shoots downy.

A culinary plum. Ripe in the middle and end of September.

WHITE IMPÉRATRICE (*Impératrice Blanche*).—Fruit medium sized, oval. Skin bright yellow, covered with very thin bloom. Stalk half an inch long, inserted in a narrow cavity. Flesh firm and transparent, juicy, sweet, and separating from the stone. Shoots smooth.

A dessert plum, requiring a wall, and ripening in the beginning and middle of September.

WHITE MAGNUM BONUM (*Askew's Golden Egg; Bonum Magnum; Dame Aubert; Dame Aubert Blanche; Egg Plum; Grosse Luisante; Impériale Blanche; White Mogul; Yellow Magnum Bonum*).—Fruit of the largest size, oval, with a rather deep suture extending the whole length of one side. Skin deep yellow, covered with thin white bloom. Stalk an inch long, inserted without depression. Flesh yellow, firm, coarse-grained, with a brisk subacid flavour, and adhering to the stone. Shoots smooth.

A culinary plum, highly esteemed for preserving. Ripe in the beginning of September.

White Mirabelle. See *Mirabelle Petite.*

White Mogul. See *White Magnum Bonum.*

WHITE PERDRIGON (*Brignole; Maître Claude; Perdrigon Blanc*).—Fruit medium sized, oval, narrowing towards the stalk, with a faint suture on one side. Skin pale yellow, strewed with white dots, and marked with a few red spots next the sun. Stalk three quarters of an inch long, slender, inserted in a small cavity. Flesh tender, juicy, rich, and slightly perfumed, separating from the stone. Shoots downy.

An excellent plum for drying and preserving. Ripe in the end of August.

WHITE PRIMORDIAN (*Amber Primordian; Avant Prune Blanche; D'Avoine; De Catalogne; Catalonian; Cerisette Blanche; Early Yellow; Jaune de Catalogne; London Plum; Pickett's July; St. Barnabe*).—Fruit small, oval, narrowing towards the stalk, marked with a shallow suture. Skin pale yellow, covered with thin white bloom. Stalk half an inch long, very slender, inserted in a small cavity. Flesh yellow, tender, sweet, and pleasantly flavoured, separating from the stone. Shoots downy.

A very early plum, but of little merit. Ripe in the end of July.

Wilmot's Early Orleans. See *Early Orleans.*

Wilmot's Green Gage. See *Green Gage.*

Wilmot's Late Orleans. See *Goliath.*

Wilmot's Orleans. See *Early Orleans.*

WINESOUR (*Rotherham*).—Fruit below medium size, oval. Skin dark purple, covered with darker purple specks. Stalk half an inch long. Flesh greenish-yellow, agreeably acid, and having red veins near the stone, to which it adheres. Shoots downy.

A very valuable preserving plum. Ripe in the middle of September.

WOOLSTON BLACK GAGE.—Fruit about medium size, round, and marked with a shallow suture. Skin deep purple, almost black, strewed with small dots, and covered with blue bloom. Flesh melting, juicy, sugary, and rich, separating from the stone. Shoots smooth.

A dessert plum of excellent quality. Ripe in the beginning of September.

Yellow Apricot. See *Apricot.*

YELLOW GAGE (*Gonne's Green Gage; Little Queen Claude; Reine Claude Petite; Petit Damas Vert; White Gage*).—Fruit below medium size, round, and marked with a shallow suture. Skin greenish-yellow, thickly covered with white bloom. Stalk half an inch long, inserted in a pretty deep cavity. Flesh yellowish-white, firm, rather coarse-grained, but sweet and pleasantly flavoured, separating from the stone. Shoots smooth.

A dessert plum of second-rate quality. Ripe in the beginning and middle of September.

YELLOW IMPÉRATRICE (*Altesse Blanche; Monsieur à Fruits Jaune*).—Fruit large, roundish-oval, marked with a suture, which is deep at the apex and becomes shallow towards the stalk. Skin deep golden yellow, with a few streaks of red about the stalk, which is half an inch long. Flesh yellow, juicy and melting, sugary and richly flavoured, and adhering to the stone. Shoots smooth.

An excellent dessert plum. Ripe in the middle of August.

Yellow Magnum Bonum. See *White Magnum Bonum.*

Yellow Perdrigon. See *Drap d'Or.*

Zwetsche. See *Quetsche.*

## LISTS OF SELECT PLUMS,

*Arranged in their order of ripening.*

### I. FOR DESSERT.

July Green Gage
Peach
De Montfort
Denniston's Superb
Perdrigon Violet Hâtif
Green Gage
Hulings' Superb
Purple Gage
Transparent Gage
Abricotée de Braunau
Jefferson
Kirke's
Topaz
Coe's Golden Drop
Reine Claude de Bavay
Cooper's Large
Late Orleans
Coe's Late Red

### II. FOR COOKING.

Early Prolific
Early Orleans
Gisborne's
Goliath
Prince of Wales
Victoria
Diamond
Autumn Compôte
Belle de Septembre

### III. FOR PRESERVING.

Green Gage
White Magnum Bonum
Diamond
Washington
Winesour
Damson
Autumn Compôte

### IV. FOR WALLS.

July Green Gage
De Montfort
Green Gage
Purple Gage
Italian Quetsche
Coe's Golden Drop
Blue Impératrice
Ickworth Impératrice

### V. FOR ORCHARDS AND MARKETING.

Early Prolific
Early Orleans
Gisborne's
Orleans
Prince of Wales
Victoria
Pond's Seedling
Damson
Coe's Late Red

# RASPBERRIES.

## SYNOPSIS OF RASPBERRIES.

### I. SUMMER BEARERS.

#### 1. *Fruit Black.*

Black
Black Cap

#### 2. *Fruit Red.*

Barnet
Carter's Prolific
Cornwell's Victoria
Cushing
Fastolf
Franconia
Knevett's Giant
Northumberland Fillbasket
Prince of Wales
Red Antwerp
Round Antwerp
Vice-President French
Walker's Dulcis

#### 3. *Fruit Yellow.*

Brinckle's Orange
Magnum Bonum
Sweet Yellow Antwerp
Yellow Antwerp

### II. AUTUMNAL BEARERS.

#### 1. *Fruit Black.*

Autumn Black
New Rochelle
Ohio Everbearing

#### 2. *Fruit Red.*

Belle de Fontenay
Large Monthly
October Red
Rogers' Victoria

#### 3. *Fruit Yellow.*

October Yellow

---

À Gros Fruits Rouges. See *Red Antwerp.*

American Black. See *Black Cap.*

D'Anvers à Fruits Ronds. See *Round Antwerp.*

AUTUMN BLACK.—This is a variety raised by Mr. Rivers from the new race of Black Raspberries which he has for some years been experimenting upon. These Black Raspberries are evidently the result of a cross between the Blackberry and the Raspberry, possessing the rambling growth of the former with the large succulent fruit of the latter. The Autumn Black produces from its summer shoots a full crop of medium-sized dark fruit of the colour of the Blackberry, and partaking much of its flavour. Ripe in October.

BARNET (*Barnet Cane; Cornwell's Prolific; Cornwell's Seedling; Large Red; Lord Exmouth's*).—The fruit is large, roundish-ovate, of a bright purplish-red colour This is larger than the Red Antwerp, but not equal to it in flavour; it is, nevertheless, an excellent variety, and an abundant summer bearer.

Barnet Cane. See *Barnet.*

BELLE DE FONTENAY (*Belle d'Orleans*).—An autumn-bearing variety of dwarf-habit, and with large leaves, quite silvery on their under surface. The fruit is large, round, of a red colour, and good flavour. Ripe in October.

The plant is a shy bearer, and throws up suckers so profusely as to be almost a weed; but if the suckers are thinned out it bears better.

Belle d'Orleans. See *Belle de Fontenay.*

BLACK.—This is a hybrid between the Blackberry and the Raspberry, and is the parent of all the black autumn-bearing varieties; although itself a summer-bearer. It has long dark-coloured canes, and small purple fruit, with much of the Blackberry flavour. This variety was obtained at Wethersfield, in Essex, upwards of forty years ago, and has since been cultivated by Mr. Rivers, who has succeeded in obtaining from it his new race of autumn-bearing black varieties.

BLACK CAP (*American Black*).—This is the Rubus occidentalis, called Black Raspberry, or Thimbleberry, by the Americans. The fruit has a fine brisk acid flavour, and is much used in America for pies and puddings. It ripens later than the other summer-bearing varieties.

BRINCKLE'S ORANGE (*Orange*).—A variety introduced from America, where it is considered the finest yellow sort in cultivation. In this country it is smaller than the Yellow Antwerp, and more acid. The plants throw up an abundance of suckers. It is a summer bearer.

Burley. See *Red Antwerp.*

CARTER'S PROLIFIC.—Fruit large and round, of a deep red colour, with a firm flesh of excellent flavour. A summer-bearing variety.

De Chili. See *Yellow Antwerp.*

Cornwell's Prolific. See *Barnet.*

Cornwell's Seedling. See *Barnet.*

CORNWELL'S VICTORIA.—The fruit of this variety is large, and of fine flavour, but its drupes adhere so closely to the core as to crumble off in gathering. A summer bearer.

CUSHING.—Fruit large, roundish, inclining to conical, of a bright crimson colour, and with a briskly-acid flavour. A summer bearer.

Cutbush's Prince of Wales. See *Prince of Wales.*

Double-Bearing Yellow. See *Yellow Antwerp.*

FASTOLF (*Filby*).—Fruit large, roundish-conical, bright purplish red, and of excellent flavour. A summer bearer.

Filby. See *Fastolf.*

FRANCONIA.—Fruit large, obtuse-conical, of a dark purplish-red colour, and good flavour, briskly acid. A summer bearer.

French. See *Vice-President French.*

Howland's Red Antwerp. See *Red Antwerp.*

Knevett's Antwerp. See *Red Antwerp.*

KNEVETT'S GIANT.—Fruit large, obtuse-conical, deep red, and of good flavour. A summer bearer.

LARGE MONTHLY (*Large-fruited Monthly; Rivers' Monthly; de Tous le Mois à Gros Fruits Rouges*).—This is a most abundant bearing autumnal variety, producing fruit above the medium size, roundish-conical, of a crimson colour, and of excellent flavour.

Large Red. See *Barnet.*

Late-bearing Antwerp. See *Red Antwerp.*

Lawton. See *New Rochelle.*

Lord Exmouth's. See *Barnet.*

MAGNUM BONUM.—A yellow summer-bearing variety, inferior in size and flavour to Yellow Antwerp. The fruit is of a pale yellow colour with firm flesh. The plant, like Brinckle's Orange and Belle de Fontenay, becomes a perfect weed from the profusion of suckers it throws up.

Merveille de Quatre Saisons Jaune. See *October Yellow.*

Merveille de Quatre Saisons Rouge. See *October Red.*

New Rochelle (*Lawton; Seacor's Mammoth*).—An American autumn-bearing variety, having the rambling habit of growth of the common Bramble. It produces fruit in great abundance of a large oval shape, and a deep black colour, very juicy, and agreeably flavoured.

This has not been sufficiently proved in this country to admit of a correct estimate being formed of its merits.

Northumberland Fillbasket.—Fruit rather large, roundish, inclining to conical, of a deep red colour, and good flavour. The plant is a strong vigorous grower, and an abundant summer bearer.

October Red (*Merveille de Quatre Saisons Rouge.*)—The fruit of this variety produced from the old canes left in spring is small and inferior; but the suckers put forth in June furnish an abundant crop of large-sized bright red fruit, which commences to ripen in September and continuing far into November, if the autumn be dry and mild.

October Yellow (*Merveille de Quatre Saisons Jaune*).—This possesses the same qualities as the preceding, and is distinguished from it by the fruit being yellow. It is not quite so large as the Yellow Antwerp, and in a fine season is sweet and agreeable.

Ohio Everbearing.—This is an American variety, similar in all respects to Black Cap, with this exception that it is an autumnal-bearing variety, and produces abundant crops of fruit late in the season.

Orange. See *Brinckle's Orange.*

Prince of Wales (*Cutbush's Prince of Wales*).—Fruit large, roundish, inclining to conical, of a deep crimson colour, and with a brisk, agreeable flavour. This is a summer-bearing variety, remarkable for its strong pale-coloured canes, which in rich soils grow from ten to twelve feet in one season. It does not sucker too much, and is very desirable on that account.

Red Antwerp (*Burley; à Gros Fruits Rouges; How-*

*land's Red Antwerp; Knevett's Antwerp; Late Bearing Antwerp*).—Fruit large, roundish, inclining to conical, of a deep crimson colour, very fleshy, and with a fine brisk flavour and fine bouquet. There are several forms of this variety differing more or less from each other both in the fruit and the canes. The true old Red Antwerp produces vigorous canes, which are almost smooth.

Rivers' Monthly. See *Large Monthly.*

ROGERS' VICTORIA (*Victoria*).—This is an autumnal-bearing variety, producing rather large, dark-red fruit of excellent flavour, and earlier than the October Red. The plant is of a dwarf and rather delicate habit, and the canes are dark coloured.

ROUND ANTWERP (*d'Anvers à Fruits Ronds*).—Fruit large and round, of a deep red colour, and much superior in flavour to the Old Red Antwerp.

Seacor's Mammoth. See *New Rochelle.*

SWEET YELLOW ANTWERP.—The fruit of this variety is larger and more orange than the Yellow Antwerp, and is the richest and sweetest of all the varieties. The canes are remarkably slender, and with few spines.

VICE-PRESIDENT FRENCH (*French*). — Fruit large, roundish, inclining to conical, of a deep red colour, fleshy and juicy, and with an excellent flavour. It is a summer bearer, producing very strong canes of a bright brown colour.

Victoria. See *Rogers' Victoria.*

WALKER'S DULCIS.—A summer-bearing variety, producing red fruit inferior in size to the Antwerp, and not sweet, as the name implies.

White Antwerp. See *Yellow Antwerp.*

YELLOW ANTWERP (*De Chili; Double-bearing Yellow; White Antwerp*).—Fruit large, conical, of a pale yellow colour, and with a fine, mild, sweet flavour. It produces pale-coloured spiny canes.

---

## LIST OF SELECT RASPBERRIES.

Autumn Black
Carter's Prolific
Fastolf
October Red
October Yellow
Rogers' Victoria
Round Antwerp
Sweet Yellow Antwerp

## STRAWBERRIES.

Aberdeen Seedling. See *Roseberry.*

Adair.—Fruit medium sized, roundish-ovate, even and regular in its shape. Skin of a uniform dark red colour. Seeds not deeply embedded. Flesh deep red throughout, rather soft and woolly, hollow at the core, not richly flavoured.

Admiral Dundas.—Fruit very large, roundish, inclining to conical, irregular and angular, sometimes cockscomb shaped; the smaller fruit conical. Skin pale scarlet. Flesh firm, juicy, brisk, and highly flavoured.

This is the best of all the very large strawberries, and was raised by Mr. Myatt.

Ajax.—Fruit large, irregularly-roundish, very deeply furrowed. Seeds deeply embedded, with prominent ridges between them, which give the surface a coarse appearance. Skin dull brick-red. Flesh deep red, and solid throughout, juicy, briskly flavoured, and tolerably rich.

The plant is of a luxuriant habit, and bears badly in the open ground; but when grown in pots it produces an abundance of fruit, and is a good forcer.

Alice Maude. See *Princess Alice Maude.*

Belle Bordelaise. Somewhat similar to *Prolific Hautbois.*

Bicton Pine —Fruit large, roundish and even in its outline. Skin pale yellowish-white, sometimes faintly tinged with red next the sun. Flesh tender and soft, juicy, brisk, and with a pine flavour.

Black Pine. See *Old Pine.*

Black Prince (*Cuthill's Black Prince*).—Fruit small, obovate. Skin glossy, of a dark red colour, which, when the fruit is highly ripened, becomes almost black. Seeds rather prominent. Flesh deep orange, brisk, rather rich, and with a little of the pine flavour.

A very early strawberry, a great bearer, and well adapted for forcing.

British Queen (*Myatt's British Queen*).—Fruit large, sometimes very large, roundish, flattened, and cockscomb shaped, the smaller fruit ovate or conical. Skin pale red, colouring unequally, being frequently white or greenish-white at the apex. Flesh white, firm, juicy, and with a remarkably rich and exquisite flavour.

The great fault of this variety is that the plant is so very tender; it will not succeed in all soils and situations, and it is generally an indifferent bearer.

Captain Cook.—Fruit large, roundish-ovate, and irregular. Skin deep scarlet, and frequently greenish at the point. Flesh pale scarlet, solid throughout, juicy and richly flavoured, but not of first-rate quality.

Carolina. See *Old Pine*.

Carolina Superba.—Fruit very large, ovate, sometimes inclining to cockscomb shape, with an even surface. Seeds not deeply embedded. Skin pale red, extending equally over the whole fruit. Flesh clear white, very firm and solid, with a fine vinous flavour and rich aroma, equalling the British Queen.

The plant is much hardier, a freer grower, and better bearer than British Queen.

Comte de Paris.—Fruit large, obtuse-heartshaped, even in its outline. Skin scarlet, becoming deep crimson when highly ripened. Flesh pale red, and solid throughout, with a briskly acid flavour.

This is a favourite with those who prefer a brisk fruit; and it is an excellent bearer.

Crimson Queen (*Doubleday's No.* 2).—Fruit large, cockscomb shape, very much corrugated and irregular, with a coarse surface. Skin bright cherry-scarlet. Flesh red throughout, solid, and firm, with a briskly acid flavour.

This is a late variety, and a great bearer.

Cuthill's Black Prince. See *Black Prince*.

Cuthill's Prince of Wales.—Fruit medium sized, conical. Skin bright red. Flesh firm, very acid, and without much flavour.

Cuthill's Princess Royal. See *Princess Royal of England*.

Deptford Pine.—Fruit large, and cockscomb shaped, the smaller fruit conical. Skin bright scarlet, glossy as if varnished, and even. Flesh scarlet, firm, and solid throughout, with a rich vinous flavour, similar to British Queen, with a little more acid.

A valuable firm-fleshed, highly-flavoured strawberry Excellent for preserving.

Downton. See *Downton Pine.*

Downton Pine (*Downton*).—Fruit medium sized, conical, with an even surface. Skin deep scarlet. Seeds embedded. Flesh scarlet, firm, and solid throughout, briskly and richly flavoured.

Doubleday's No. 2. See *Crimson Queen.*

Duchesse de Trévise (*Marquise de la Tour Maubourg; Vicomtesse Héricart de Thury*).—Fruit above medium size, conical, with an even surface. Skin deep scarlet, becoming deep red as it ripens. Seeds yellow, slightly embedded. Flesh pale red throughout, firm and solid, brisk, sweet, and richly flavoured.

This is an extraordinarily abundant bearer, and a valuable variety for general cultivation.

Eleanor (*Myatt's Eleanor*).—Fruit very large, conical or wedge-shaped, regular and handsome in its outline. Seeds considerably embedded, with prominent ridges between them, which give the fruit a coarse appearance on the surface. Skin scarlet, changing as it ripens to deep crimson. Flesh scarlet, and becoming paler towards the core, which is large and hollow; subacid, and with a little of the pine flavour.

A large and handsome strawberry, but not possessing any other merit.

Eliza. See *Myatt's Eliza.*

Elton (*Elton Pine*).—Fruit large, ovate, frequently cockscomb shaped, with embedded seeds, and prominent ridges between them. Skin bright crimson, and shining. Flesh red throughout, firm and solid, with a brisk subacid flavour.

A valuable late variety, and an excellent bearer.

Elton Pine. See *Elton.*

Empress Eugénie.—Fruit very large, irregular, angular, furrowed, and uneven. Skin of a deep red colour, becoming almost black when highly ripened. Seeds small, not deeply embedded. Flesh red throughout, hollow at the core, tender, very juicy, and briskly flavoured.

Rather a coarse-looking and very large strawberry, not remarkable for any excellency of flavour.

Exhibition. See *Great Exhibition.*

FILBERT PINE (*Myatt's Seedling*).—Fruit above medium size, conical and regular in its outline. Seeds large and prominent. Skin dull purplish-red next the sun and pale red in the shade. Flesh pale, pink at the core, firm, solid, rich, and briskly flavoured, with a fine aroma.

A very prolific and excellent late variety.

FILLBASKET.—Fruit rather large, roundish, sometimes flattened on the sides. Skin dark red. Flesh pale red throughout, very acid, and without much flavour.

Goliath. See *Kitley's Goliath.*

GREAT EXHIBITION (*Exhibition*).—Fruit medium sized, oblong, ovate, or irregular. Seeds prominent. Skin bright red. Flesh dull yellow, very woolly and worthless.

The plant is a great bearer, but otherwise not worth growing.

HIGHLAND CHIEF.—Fruit large, roundish-ovate, and somewhat flattened. Seeds not deeply embedded. Skin fine, clear red, becoming darker red as it ripens. Flesh dark red throughout, very firm and solid, very juicy and vinous, and with a rich pine flavour.

A very excellent strawberry. The plant is a most abundant bearer, and deserves universal cultivation.

HOOPER'S SEEDLING.—Fruit large, conical, rarely flattened, but sometimes deeply furrowed. Seeds rather deeply embedded. Skin dark red, assuming a very deep blackish tinge as it ripens. Flesh crimson at the exterior, but paler towards the centre, sweet, brisk, and richly flavoured.

A good bearer, and an excellent variety for general purposes.

INGRAM'S PRINCE ARTHUR.—Fruit medium sized, conical, even and regular in shape, with a glossy neck. Seeds not very numerous, nor deeply embedded. Skin of a brilliant scarlet, like Sir Charles Napier, paler at the tip. Flesh white, solid, very juicy, brisk, and with a rich pine flavour.

A first-rate variety, an abundant bearer, and forces well.

Ingram's Prince of Wales. — Fruit very large, roundish, flattened and wedge-shaped, the smaller fruit ovate. Seeds not deeply embedded. Skin deep crimson, becoming darker as it ripens. Flesh pale red, very firm and solid, brisk, sweet, and richly flavoured.

An excellent variety, and admirably adapted for forcing.

Keens' Seedling.—Fruit large, ovate, sometimes inclining to cockscomb shape. Seeds not deeply embedded. Skin dark crimson, becoming very dark when highly ripened. Flesh scarlet, firm and solid, juicy, brisk, and richly flavoured.

An old and well-established variety, which, for general purposes, has not yet been surpassed.

Kitley's Goliath (*Goliath*).—Fruit very large, compressed and wedge-shaped, the smaller ones ovate. Seeds deeply embedded, which gives the surface a rough appearance. Skin deep red, colouring equally all over. Flesh white, solid, briskly and richly flavoured, but not equal to British Queen, to which it is similar. It is, however, a better grower and better cropper.

Mammoth (*Myatt's Mammoth*). — Fruit immensely large, flattened, deeply furrowed and ribbed, irregular and uneven in its outline. Seeds small and very slightly embedded. Skin glossy, of a fine deep red colour. Flesh scarlet throughout, firm and solid, even in the largest specimens, and of a brisk and pleasant flavour, which is rich in the well ripened fruit.

The foliage is small, and on short footstalks, and permits the fruit to be well exposed to the influence of the sun.

Marquise de la Tour Maubourg. See *Duchesse de Trévise.*

Myatt's British Queen. See *British Queen.*

Myatt's Eleanor. See *Eleanor.*

Myatt's Eliza.—Fruit above medium size, ovate or conical, with a glossy neck. Seeds not deeply embedded. Skin light red, becoming deep red when highly ripened. Flesh scarlet on the outside, but paler towards the core, firm and solid, very juicy, and with a particularly rich and exquisite flavour.

This is one of the richest flavoured of all the varieties. The plant is a pretty good bearer, and hardier than the British Queen, to which it is, under all circumstances, superior in flavour.

Myatt's Globe.—Fruit large, roundish-ovate, even and regular, and with rather prominent seeds. Skin pale red, or rose coloured. Flesh white, but not solid at the core, of a rich and excellent flavour.

The plants are most abundant bearers.

Myatt's Mammoth. See *Mammoth.*

Myatt's Seedling. See *Filbert Pine.*

Myatt's Surprise. See *Surprise.*

Ne Plus Ultra.—Fruit large, cylindrical or oblong, frequently assuming a digitate shape. Skin very dark red. Flesh remarkably firm and solid, with a rich and pleasant flavour.

This is a singular variety, many of the fruit being so divided at the apex as to appear like fingers.

Nimrod.—I have not yet been able to meet with what is said to be the true form of this variety, all the plants I have seen in fruit having proved to be the same as *Eleanor.*

Old Pine (*Black Pine; Carolina; Scarlet Pine*).—Fruit medium sized, ovate, even and regular, and with a glossy neck. Seeds prominent. Skin deep red. Flesh pale red, very firm and solid, with a fine sprightly and very rich pine flavour.

After all there are very few that equal, far less surpass, the Old Pine in flavour; but it is not a good bearer.

Omar Pasha (*Rival Queen*).—Wherever I have met with this variety it has proved to be the same as *Myatt's Eliza.*

Oscar.—Fruit large, ovate and angular, sometimes flattened and wedge-shaped. Seeds rather large, and deeply embedded, which give the surface a coarse appearance. Skin dark shining red, becoming almost black when fully ripe. Flesh red throughout, very firm and solid, juicy and richly flavoured.

An excellent variety for a general crop, coming in a few days after Black Prince; a most abundant bearer; and from its firmness bears carriage well.

Princess Alice Maude (*Alice Maude*).—Fruit medium sized, ovate or conical, and frequently large and kidney-shaped. Seeds prominent or very slightly embedded. Skin scarlet, becoming dark crimson when ripe. Flesh scarlet throughout, tender, juicy, sweet, and with a rich, brisk flavour.

PRINCESS ROYAL OF ENGLAND (*Cuthill's Princess Royal*).—Fruit medium sized, roundish-ovate or conical, with a neck. Seeds deeply embedded. Skin deep scarlet, where exposed to the sun, and paler in the shade. Flesh pale red at the surface, whitish towards the core, very rich and highly flavoured.

An abundant bearer, and an excellent variety for general cultivation.

PROLIFIC HAUTBOIS.—Fruit below medium size, conical. Seeds prominent. Skin light purple in the shade, and blackish-purple on the side next the sun. Flesh firm, sweet, and with the rich peculiar flavour of the Hautbois.

Prolific Pine. See *Roseberry.*

Rival Queen. See *Omar Pasha.*

RIVERS' ELIZA (*Seedling Eliza*).—This is a seedling from Myatt's Eliza, but rather more ovate in shape, and possessing all the character and flavour of that excellent variety, but is a more abundant bearer, and of a hardier constitution.

ROSEBERRY (*Aberdeen Seedling; Prolific Pine*).—Fruit large, conical, and pointed. Seeds deeply embedded, with prominent ridges between them. Skin dark red, becoming blackish as it ripens. Flesh pale scarlet, firm, with an agreeable flavour.

Royal Pine. See *Swainstone's Seedling.*

RUBY—Fruit large, roundish, dark red. Flesh pale red, soft, and woolly, with a large core, and inferior flavour.

Scarlet Pine. See *Old Pine.*

Seedling Eliza. See *Rivers' Eliza.*

SIR CHARLES NAPIER.—Fruit very large, ovate, flattened, and wedge-shaped. Seeds not deeply embedded. Skin shining, of a fine bright pale scarlet colour. Flesh white, firm and solid, briskly acid, and not highly flavoured.

This is a fine handsome strawberry, well adapted for forcing and for early market purposes. The plant is remarkably tender, perhaps more so than any other variety.

SIR HARRY.—Fruit very large, roundish, irregular, frequently cockscomb-shaped. Seeds large, and deeply embedded. Skin dark crimson, becoming almost black when fully ripe. Flesh dark red, not very firm, but tender, very juicy, and richly flavoured.

Sir Walter Scott.—Fruit medium sized, conical, and pointed, with prominent seeds. Skin deep red. Flesh pale, firm, and inferior in flavour.

Stirling Castle Pine.—Fruit large, ovate or conical, pointed, even and regular in shape. Seeds small, not deeply embedded. Skin of a bright scarlet colour, becoming dark red as it ripens. Flesh pale scarlet, brisk, and of excellent flavour.

Swainstone's Seedling (*Royal Pine*).—Fruit above medium size, ovate, even and regular in its shape. Seeds small, and rather deeply embedded. Skin pale red. Flesh pale, rather hollow round the core, and with a fine rich flavour.

This is a good variety for forcing, and is a good bearer.

Trollope's Victoria.—Fruit very large, roundish-ovate, even and regular in its outline. Skin light crimson. Flesh pale scarlet, tender, juicy, sweet, and richly flavoured.

This is a good early strawberry, and an excellent bearer.

Viscomtesse Héricart de Thury. See *Duchesse de Trévise.*

Wilmot's Prince Arthur.—Fruit medium sized, conical, even, and regular. Seeds small, not deeply embedded. Skin deep red and glossy. Flesh scarlet, firm, but hollow at the core, of a rich flavour when highly ripened.

The plant is a great bearer, forces well, and the fruit bears carriage better than many other varieties.

---

## LIST OF SELECT STRAWBERRIES.

Black Prince
British Queen
Carolina Superba
Deptford Pine
Duchesse de Trévise
Elton
Highland Chief
Keens' Seedling
Myatt's Eliza
Oscar
Princess Royal of England
Swainstone's Seedling

## WALNUTS.

A Bijoux. See *Large Fruited.*

Common.—The common walnut being raised from seeds there are a great number of varieties among those grown in this country, varying in size, flavour, thickness of the shell, and fertility. To secure a variety of a certain character, it must be perpetuated by grafting in the same way as varieties of other fruit trees are propagated.

À Coque Tendre. See *Thin Shelled.*

Double. See *Large Fruited.*

Dwarf Prolific (*Early Bearing; Fertile; Præparturiens; Precocious*).—This is a dwarf-growing, early-bearing variety, which I have seen produce fruit when not more than two and a half to three feet high; and a tree in my possession, not more than six feet high, bears abundant crops of good-sized and well-flavoured fruit. This variety reproduces itself from seed.

Early Bearing. See *Dwarf Prolific.*

Fertile. See *Dwarf Prolific.*

French. See *Large Fruited.*

Highflyer.—This variety ripens its fruit considerably earlier than the others, and is of good size and well flavoured.

De Jauge. See *Large Fruited.*

Large Fruited (*à Bijoux; Double; French; de Jauge; à Très Gros Fruit*).—Nuts very large, two or three times larger than the common walnut, and somewhat square or oblong in shape. The kernel is small for the size of the nut, and does not nearly fill the shell. It requires to be eaten when fresh, as it very soon becomes rancid.

The shell of this variety is used by the jewellers for jewel-cases, and is frequently fitted up with ladies' embroidery instruments.

Late (*Tardif; Saint Jean*).—The leaves and flowers of

this variety are not developed till near the end of June, after all danger from frosts has passed. The nuts are of medium size, roundish, and well filled; but they do not keep long. The tree is very productive, and is reproduced from the seed.

À Mésange. See *Thin Shelled.*

Præparturiens. See *Dwarf Prolific.*

Precocious. See *Dwarf Prolific.*

Saint Jean. See *Late.*

Tardif. See *Late.*

THIN SHELLED (*à Coque Tendre; à Mésange*).—Nuts oblong, with a tender shell, and well filled. This is the best of all the varieties.

À Très Gros Fruit. See *Large Fruited.*

YORKSHIRE.—This is of large size, but not so large as the Large Fruited. It fills and ripens well.

# SUPPLEMENT.

—o—

## APPLES.

Baron Ward.—Fruit below medium size, ovate. Skin smooth and shining, of a fine uniform deep yellow colour. Eye slightly open, and not much depressed. Stalk short. Flesh tender, crisp, juicy, and agreeably acid. January till May.

This is an excellent apple for culinary purposes, but its small size is a great objection to it. It keeps well without shrivelling.

Clissold's Seedling. See *Lodgemore Nonpareil.*

Duke of Devonshire.—Fruit medium sized, roundish-ovate. Skin of uniform lemon-yellow colour, with a dull red cheek; the surface veined with russet. Eye large and open, set in a wide and deep basin. Stalk very short. Flesh yellowish, crisp, juicy, rich, and sugary, with a fine aroma.

An excellent dessert apple, in use from February till May.

Lodgemore Nonpareil (*Clissold's Seedling*).—Fruit about medium size, roundish. Skin deep yellow, dotted with minute grey dots, and with a blush of red on one side. Eye slightly closed, set in a shallow basin. Stalk short, deeply inserted in the cavity. Flesh yellowish, firm, crisp, juicy, and sugary, with a fine aroma.

A first-rate dessert apple, in use from February till the beginning of June.

## APRICOTS.

Canino Grosso.—This is a fine large apricot, ripening at the same time as Royal; remarkably robust in its habit of growth, and likely to prove a desirable sort; but it has not been sufficiently proved in this country to know what its real merits are.

Précoce de Wittemberg.—This is an early variety of the Peach Apricot, and as such is highly valuable. It ripens ten or twelve days before that variety, and is of the largest size.

## GOOSEBERRIES.

Companion.—Large, roundish-oval. Skin hairy, of a pale red colour in the shade, and brownish-red next the sun. Flavour very rich and excellent.

This is one of the best gooseberries, combining size and flavour. Its greatest weight is 26 dwts. 8 grs.

Freedom.—Large and oval. Skin thin, smooth, greenish-white, with streaks of red on the side next the sun. Flavour sweet and good. Greatest weight 22 dwts. 22 grs.

Leader.—Large and round. Skin yellow, rather thick and smooth. Flavour excellent. Greatest weight 23 dwts. 20 grs.

Lion's Provider.—Large and oval. Skin rather thick, hairy, dark red, and somewhat transparent. Flavour sweet and brisk. Greatest weight 25 dwts. 8 grs.

Wonderful.—Large and oval. Skin smooth, rather thin, transparent, and deep red. Flavour rich. Greatest weight 28 dwts. 12 grs.

## GRAPES.

Black Monukka.—Bunches very large, and well set; ovate, and broadly shouldered. Berries of an oblong-ovate shape, like those of the Finger Grape, dark red or nearly black, and set on long slender stalks, which are very brittle. Skin very thin, adhering so closely to the flesh as to be inseparable when the fruit is eaten. Flesh very firm and crisp, juicy, sweet, and nicely flavoured. The berries are stoneless.

This is a strong, vigorous-growing vine, and very productive.

Duc de Malakoff (*Chasselas Duc de Malakoff*).—This is a form of the Sweetwater, and in all respects so nearly resembles that variety that it is not worth keeping distinct. From what I have seen of it, it sets as badly as the Sweetwater, and produces a bunch with a few large and a great many small berries.

Early Green Madeira (*Vert Précoce de Madère*).—Bunches of good size, cylindrical, slightly compact. Berries medium sized, oval. Skin of a green colour, which it retains till its perfect maturity, when it becomes a little clearer, but still preserving the green tinge. Flesh with a rich and sugary flavour.

This is one of the earliest grapes, and ripens in a cool vinery from the beginning to the middle of August. It will also succeed against a wall in the open air; but, of course, is not then so early. It bears considerable resemblance to the Verdelho, but is said to be earlier than that variety. I have not been able to examine the two growing under the same circumstances.

Ingram's Hardy Prolific Muscat.—Bunches long and tapering, not shouldered, from nine inches to a foot in length. Berries medium size, perfectly oval, and well set. Skin quite black, covered with blue bloom. Flesh moderately firm, juicy, sugary, and with a fine piquant and rich flavour, having a faint trace of Muscat.

This is an excellent hardy grape, and remarkably prolific. The wood is very short-jointed, and the vine succeeds well in a cool greenhouse. It has all the appearance of being a good out-door grape.

Jura Frontignan (*Muscat Noir de Jura*).—Bunches long and tapering, very slightly shouldered, and larger than those of Black Frontignan. Berries above medium size, oval, and well set. Skin deep purplish-black, covered with thin blue bloom. Flesh tender, very juicy, richly flavoured, and with a fine, but not powerful, Muscat aroma.

This is a valuable grape. The vine is a prolific bearer;

the wood short-jointed; and will be well adapted for growing in pots.

MILL-HILL HAMBURGH.—Since the preceding portion of this work has passed through the press I have had new opportunities of examining the characters of this grape, and in addition to the distinction of foliage I find there are other differences to separate it from the Dutch Hamburgh than those mentioned at page 109. The Dutch Hamburgh has a firm, coarse flesh adhering to the skin; but that of the Mill-Hill is perfectly tender, and both in texture and flavour resembles the true Black Hamburgh, while the berries are as large and of the same shape as those of Dutch Hamburgh. It is a very fine variety, and perfectly distinct.

MUSCAT CITRONELLE.—Bunches small, and not shouldered. Berries below medium size, like those of Royal Muscadine, round. Skin thin, and somewhat transparent, white, and covered with thin bloom. Flesh very tender, juicy, and sweet, with a slight Muscat flavour. An early grape, ripening in a cool greenhouse in the middle of August.

VERDAL.—Bunches long, loose, and tapering, not shouldered. Berries above medium size, oval, on long slender stalks. Skin thin, green, covered with thin bloom. Flesh tender, very juicy, sweet, and richly flavoured.

This is an excellent early grape, ripening in a cool vinery in the middle of August.

## PLUMS.

MITCHELSON'S.—Fruit above medium size, oval, not marked with a suture on the side. Skin black when fully ripe, dotted with a few very minute fawn-coloured dots, and covered with a very thin blue bloom. Stalk half an inch long, stout, and inserted in a depression. Flesh yellow, tender, very juicy, sweet, and of good flavour, separating from the stone. Shoots smooth.

An excellent preserving plum. Ripe in the beginning of September. In general appearance it is like the Diamond, but smaller, and does not possess that very brisk acidity which characterises that variety. It is a prodigious bearer, the fruit being produced in clusters, and it is invaluable as a market plum.

Oullens' Gage (*Reine Claude d'Oullens; Reine Claude Précoce*).—Fruit not so large as the Green Gage, but of the same shape. When ripe the skin is of a rich yellow colour, dotted with crimson on the side exposed to the sun, and covered with a very delicate white bloom. Stalk three quarters of an inch long, inserted in a rather wide depression. Flesh yellow, very tender and juicy, rich, sugary, and delicious, adhering slightly to the stone. Shoots smooth. Ripe in the middle of August.

This is a remarkably fine dessert plum, and valuable for its earliness. The tree has a robust pyramidal growth.

## STRAWBERRIES.

Culverwell's Sanspareil.—Fruit long and tapering, rarely assuming any other shape; very much furrowed and irregular on its surface. Seeds not deeply embedded. Skin very dark red, becoming almost black when highly ripened. Flesh very firm and solid, red throughout, and very richly flavoured.

Frogmore Late Pine.—Fruit very large, conical, and cockscomb-shaped, with a glossy neck like the old Pine Seeds not deeply embedded. Skin glossy, bright red. becoming dark red and almost black when ripe. Flesh tender, and very juicy, red throughout, richly flavoured, and a good deal of the Pine aroma when well ripened.

This is a late variety, and an abundant bearer, coming in with the Elton, but much less acid that that variety.

Highland Mary.—Fruit above medium size, conical, and inclining to cockscomb shape. Skin dark red. Seeds small, not deeply embedded. Flesh white, rather hollow at the core, briskly and agreeably flavoured.

The plant is an abundant bearer.

Richard the Second.—This is an improved variety of Black Prince, to which it is similar in form and colour, but of a larger size. It is above medium size, and almost round; of a dark red colour, and almost black when highly ripened. Flesh pale scarlet, firm, but hollow round the core.

Printed in the United States
By Bookmasters